Synthesis Lectures on Global Engineering

Series Editor

Gary Lee Downey, Blacksburg, VA, USA

This series offers learning about engineers, engineering practices, and engineering outcomes within and across the borders of countries. Readers gain increased abilities to engage forms of engineering knowledge, identities, and commitments that have emerged and traveled across different spaces and times. Engineering students and working engineers learn to better reflect critically on their own knowledge, identities, and commitments in relation to those of other engineering and non-engineering practitioners. Projects in the series often aim to engage variable audiences within and beyond their academic fields, thus helping those fields to expand their reach through practices of knowledge production, expression, and travel.

Kuo-Hui Chang · Gary Lee Downey · Bono Po-Jen Shih

Engineers and the Two Taiwans

The Abnormal Club

Springer

Kuo-Hui Chang
National Taiwan University
Taipei, Taiwan

Gary Lee Downey
Virginia Tech
Blacksburg, VA, USA

Bono Po-Jen Shih
Pennsylvania State University
University Park, PA, USA

ISSN 2160-7664 ISSN 2160-7672 (electronic)
Synthesis Lectures on Global Engineering
ISBN 978-3-031-59765-7 ISBN 978-3-031-59766-4 (eBook)
https://doi.org/10.1007/978-3-031-59766-4

This Springer imprint is published by the registered company Springer Nature Switzerland AG
The registered company address is: Gewerbestrasse 11, 6330 Cham, Switzerland

Engineers and the Two Taiwans:

The Abnormal Club

Departing from Taiwan (要離開台灣)

Artist: 陳秀芳 Susan Hsiu-Fang Liu, January 2013

Courtesy of Susan Hsiu-Fang Liu and Taiwanese American Archives

To the engineers and technicians who contribute themselves and their knowledge to make and re-make Taiwan

Acknowledgements

This project has received enormous support from many directions. Our main source of research support has been Taiwan's National Science and Technology Council, enabling us to visit and observe critical sites, travel to conduct interviews, transcribe and translate interviews, visit archives and collect documents and images, and carry out innumerable other essential tasks. Also, funding from the Graduate Institute of National Development at National Taiwan University; the Department of Science, Technology, and Society at Virginia Tech; and the Leonhard Center for Enhancement of Engineering Education at Pennsylvania State University helped us to allocate the time necessary to carry out the research and complete the writing.

We are grateful to the many engineers and administrators who openly shared with us their knowledge and insights, including Chew Tsai-Hsin, Chi Gou-Chung, Chiu Chun-Pang, Rockwell Hsu, Huang Kuan-Tsae, Ku Po-Kwan, Roger Liao, Lin Mou-Shiung, Wu Chin-Ju, Yang Chi-Hang (Joseph), Yeh Chiunn-Shyong (Eugene), Yu John Pang, Chang Ching-Liang, Wang Hwan-San (Samuel), Mandy Liu, Yang Yeong-bin, Yen Jia-Yush, and others who must remain anonymous. We thank the following organizations for crucial assistance in identifying and obtaining important data and materials: Academia Historica, Computer History Museum, Industrial Technology Research Institute, Executive Yuan, National Archives Administration, National Central Library, National Library of Public Information, Pan Wen-Yuan Foundation, Sun Yun-Suan Memorial Museum, Land Reform Museum, National Palace Museum, Academia Sinica, National Yang Ming Chiao Tung University, National Cheng Kung University, National Taiwan University, Taiwanese-American Archives, Institute of Engineering Education Taiwan, and North America Taiwanese Engineering & Science Association.

We have benefitted greatly from the comments and suggestions of many colleagues as we presented portions of this manuscript at the Society for Social Studies of Science, Society for History of Technology, International Society for the History of East Asian Science Technology and Medicine, and Taiwan STS Association, as well as the University of Science and Technology and Jeonbuk National University in South Korea and National Yang Ming Chiao Tung University in Taiwan.

Kuo-Hui: My department, the Graduate Institute of National Development at National Taiwan University, generously provided me with resources and space to enable me to research and write this book. I greatly appreciate my many fruitful conversations about it with faculty and student colleagues. My co-authors were my fundamental base to keep me determined and help me get through the relentless challenges we faced. They inspired me with their impressive perspectives and insights. I owe a huge debt to them, especially Gary Downey. I deeply realize my academic career would have suffered to a degree I could not imagine if I had not worked with him. He is also a precious friend from whom you can learn and benefit from his wisdom. In addition, I would like to thank Kuo Wen-Hua, Jongmin Lee, and Manyong Moon for their encouragement and advice regarding this book project.

Lastly, I am very grateful for the wonderful family Mei-Chen, Yi-Kuan, and I have created. You have given me the most valuable gifts of my life. Any accomplishments I have achieved have been because of you. Mei-Chen, I dedicate my contribution to this book to you. You have walked with me every step of the way with your wisdom and inspiration.

Gary: I am grateful for having had the opportunity to collaborate with two fine scholars and close friends, both of whom and whose families became my teachers. I have long been amazed by Kuo-Hui's vast knowledge of the technical and policy landscapes of Taiwan, his remarkable ability to embody different perspectives and sensitivity to the nuances of alternative interpretations, and his great generosity in helping others learn, including me. And I have benefitted greatly from Bono's analytical acumen, ability to articulate and (re)frame finding and argument, and insightful commitment to locating and integrating the detailed point that proves to be so much more, all owing to his remarkable trajectory through engineering to STS and back. Both of them have lasting attachments to National Taiwan University. That I was also able to gain a Taida attachment when I served as visiting Chair Professor in the Graduate Institute of National Development with the support of the National Science and Technology Council has delighted me, both enacting the material connection that lives in this collaboration and helping me personally to fulfill dreams of learning from others that I have carried throughout my career. I am grateful for the opportunity and thank Kuo-Hui in particular for making it happen.

I also thank Virginia Tech for supporting my research visit to Taiwan during my 2016 sabbatical year, and I thank the STS Department for providing years of assistantship and summer support for graduate student research on this project. I am permanently connected to the late Wu Chyuan-Yuan, my colleague and friend, who first brought me to Taiwan to lecture in 2003 with the support of National Tsing Hua University, and who organized both interviews and introductions to local industries, small and large. During these visits, I learned much from colleagues as I delivered lectures at National Tsing Hua University, National Yunlin University of Science and Technology, National Taiwan University, Academia Sinica, and National Yang Ming Chiao Tung University, and a keynote address at the Taiwan STS Association. I am privileged to have had the opportunity to know and learn from many other fine STS scholars in, around, and about Taiwan. Beginning with Lin Chung-Hsi, a key source of lasting learning for me, they have included Yu Bingbing, Lo Kuan-Hung, Lei Sean Hsiang-lin, Wu Chia-Ling, Kuo Wen-Hua, Chen Hsin-Hsing, Lin Yi-Ping, and Tsou Tsung-Yen, as well as Masanori Wada, Kyonghee Han, Brent Jesiek, and Qin Zhu.

I have experienced great joys and support from family members who have taught me much. I am grateful to Leah, Michael, Kate, Reid, Bode, Megan, Adam, Ben, Jamie, Nick, Andrew, and Casey. And special thanks to Uncle Bill Zeilfelder, my first and continuing mentor in engineering, for showing me with his example how to work with others while helping them to recognize value in themselves. I am thrilled on a daily basis to have a wonderful partner with whom I share profound trust and the joy of adventure. Marta, my contribution to this book is for you.

Bono: I thank the Department of Science, Technology, and Society at Virginia Tech for providing graduate research assistantships in Spring 2014 and Spring 2015 as well as summer employment between 2014 and 2021 to support this project. Special thanks go to the Travel Grant Committee of the Society for History of Technology enabling my conference travel to Singapore in 2016. I am also indebted to the Leonhard Center and the Rock Ethics Institute at Penn State for their support while completing this book. I greatly appreciated comments from Janet Abbate, Kari Zacharias, Joshua Earle, Tsai Hung-Yin, Samantha Fried, and Jonathan Grunert as I worked through an early course paper related to this project. Their insightful comments, constructive critiques, and encouraging feedback played a substantial role in my professional development.

Carrying out Taiwanese studies in the United States would not have been possible without access to literature originating in Taiwan. I was much obliged for the dedicated work of Sharon Gotkiewicz, Shanna Alley, Mary Lucado, Nancy Weaver, Amanda Pauley, and Vicki Cox at Virginia Tech's Interlibrary Loan services. My mom Tsai Meng-Chuan and my brother Shih Po-Hsu were my last resort when all the other means failed, including dropping off books and food throughout my mandatory 2-week home quarantine during the COVID-19 pandemic in Taiwan.

I appreciate support and encouragement from my two co-authors for the past many years while I had to be away from this book project for my doctoral and postdoc works and several life events. I first knew each of them as my professors in 2012 and 2013, and as time went on, they also became my colleagues and friends. The insight and guidance they generously offer, as well as the scholarship they each pioneer and demonstrate, span far beyond our collaboration on this book and leave an indelible imprint on the formation of my scholarship. In addition, I am grateful that my wife Felice Lu took up the slack for housework and the care of our children when I worked on this book and other projects. Her love for me and our growing family has always been a great source of strength and inspiration.

I dedicate my contribution to this book to my late father, Shih Chun-Hsien, who was a chemical engineer working diligently at Hitachi Chemical of Kaohsiung to raise his beloved small family. Remembering his decision (around the year 1967) to study chemical engineering in Kaohsiung Junior College of Technology's (高雄工專) five-year programs (which admitted junior high school graduates) instead of going to Kaohsiung High School (高雄中學) inspires me to appreciate how the evolution of engineering education and practices in Taiwan intersected with the economic development that he once went through and contributed to along with my mom and many participants of their generation.

About the Book

The identity of Taiwan is a matter of intense local and international debate. This book shows how engineers helped produce two distinct visions of the land of Taiwan and its people, elevating the value of engineering as a career path in the process.

The technical work of engineers that typically avoids partisan politics can actually contribute to partisan differences and conflict because engineers have different identities, commitments, and forms of knowledge. In most countries, the rise of engineering is about the emergence of a single dominant vision of the country. In Taiwan, two distinct visions of the land and its people emerged, owing to the presence of two distinct geographical identities—islander and mainlander.

The unique collaboration among small and large companies in the electronics industry proved to be transformative, producing an "abnormal club" of mainlander and islander engineers, as one mixed group had earlier called itself. However, because the two collections of engineers had contrasting geographical identities, their work in electronics actually aimed to produce different Taiwans.

Contents

About the Authors

Kuo-Hui Chang is Associate Professor in the Graduate Institute of National Development, National Taiwan University. A policy researcher focusing on power relations among science, technology, and the state, he completed his Ph.D. in science and technology studies at Virginia Tech following a master's degree in public policy at National Taipei University. He served as Editor-in-Chief of the *Taiwanese Journal for Studies of Science, Technology and Medicine* (STM) (2021–2022). His works appear in *Engineering Studies*, the *Korea Journal for History of Science*, and a number of Chinese journals in Taiwan, including *STM*, *Journal of Social Sciences and Philosophy*, *Journal of Public Administration*, and *Taiwan: A Radical Quarterly in Social Studies.* He is currently investigating the role of Taiwanese-American engineers in Taiwan's state transformation and technology policymaking, and he is comparing energy policy development in Taiwan with that elsewhere in East Asia.

Gary Lee Downey is an ethnographic listener committed to engineering studies, STS making & doing, and exploring/inflecting connections between knowledge and personhood. Trained as a mechanical engineer (B.S. Lehigh) and cultural anthropologist (Ph.D. Chicago), he is Alumni Distinguished Professor Emeritus of Science, Technology, and Society and affiliated faculty member emeritus in Women's and Gender Studies and Engineering Education. His books include *Making & Doing: Activating STS through Knowledge Expression and Travel* (MIT Press, 2021, co-edited with Teun Zuiderent-Jerak); *Engineers for Korea* (Morgan & Claypool, 2014, with Kyonghee Han); *What is Global Engineering Education For?* (Morgan & Claypool, 2011, co-edited with Kacey Beddoes); *The Machine in Me: An Anthropologist Sits Among Computer Engineers* (Routledge, 1998); and *Cyborgs and Citadels: Anthropological Interventions in Emerging Sciences and Technologies* (SAR Press, 1998; co-edited with Joseph Dumit). Co-founder of the International Network for Engineering Studies, he edits the Engineering Studies series at MIT Press (with Matthew Wisnioski) and the Global Engineering Series at Springer Nature. As president of the Society for Social Studies of Science (2013–2015), he co-led the creation of its initiative in STS making & doing, inflecting STS knowledge production through the scholarship of knowledge expression and knowledge travel.

Bono Po-jen Shih is an interdisciplinary scholar working in the intersection of philosophy, history, and sociology of engineering with an eye on contemporary engagement with engineering education and practice. His publications appear in Springer's *Philosophy of Engineering and Technology* (PET) book series, the journal *Techné: Research in Philosophy and Technology*, and the *Taiwanese Journal for Studies of Science, Technology and Medicine*. He currently holds a postdoc appointment with two institutions at Penn State University—the Rock Ethics Institute and the Leonhard Center for Enhancement of Engineering Education—to facilitate exchange and collaboration between philosophers and engineers. Prior to joining Penn State, he was a Postdoctoral Research Fellow at the Science History Institute working on the history of engineering ethics education. He earned his PhD and MS in science and technology studies (STS) from Virginia Tech. He also has a graduate certificate in engineering education (ENGE) from Virginia Tech and a Bachelor of Science in electrical engineering from National Taiwan University.

Abbreviations

228	Incident on February 28, 1947
31 Society	31 engineers sent by Republic of China's Resource Committee to the United States to study industrial and infrastructural engineering in 1942 (Also 31st year of Republic of China)
Academic Society	Chinese-American Academic and Professional Society
AMI	American Microsystems, Inc.
AT&T	American Telephone and Telegraph Company
CCP	Chinese Communist Party
CECI	Chinese Engineering Consultants, Inc.
Chiao Tung University	National Chiao Tung University
Chinese Institute	Chinese Institute of Engineers
Classical learning	Classical Chinese learning
CPC	Chinese Petroleum Corporation
CTCI	China Technical Consultants, Inc.
CUSA	Council for United States Aid
DPP	Democratic Progressive Party
EIRC	Electronics Industry Research Center
Fairchild	Fairchild Semiconductor International Inc.
Formosan Association	Formosan (Taiwan Provincial) Association for the Advancement of Science
High-tech Forum	U.S. Taiwan High-tech Forum
IBM	International Business Machine
IC	Integrated circuits
ICT	Information and computer technologies
Industrial Technology Association	Taiwanese-American Industrial Technology Association—Silicon Valley

ITRI	Industrial Technology Research Institute
KMT	Kuomintang (Nationalist Party)
METS	Modern Engineering and Technology Seminar
MIT	Massachusetts Institute of Technology
MOEA	Ministry of Economic Affairs
Monte Jade	Monte Jade Science and Technology Association
NATEA	North America Taiwanese Engineering & Science Association
National Council	National Science and Technology Council
National Labs	National Applied Research Laboratories
OEM	Original equipment manufacturing
PRC (People's Republic)	People's Republic of China
Professionals Association	Chinese-American Professionals Association
Professors Association	North American Taiwanese Professors Association
RCA	Radio Company of America
ROC	Republic of China
RSEA	Ret-Ser Engineering Agency
Science park	Science-based industrial parks
SMIC	Semiconductor Manufacturing International Company
SMEs	Small- and medium-sized companies/enterprises
STPI	Science and Technology Policy Research and Information Centre
STS	Science and Technology Studies
Taida	National Taiwan University
Taipei Institute	Provincial Taipei Institute of Technology
Taipei School	Industrial schools—Taipei Industrial School (Institute)
Taiwan Electric (Taiwan Power; Taipower)	Taiwan Electric Company
TSMC	Taiwan Semiconductor Manufacturing Company
TI	Texas Instruments
Tsing Hua University	National Tsing Hua University
UMC (United Microelectronics)	United Microelectronics Corporation
United Formosans	United Formosans in America for Independence

USA	United States of America
Worldwide Semiconductor	Worldwide Semiconductor Manufacturing Company
Yushan	Monte Jade Science and Technology Association in China

List of Figures

1 Abnormal Engineers for an Abnormal Land

They were both members of the Abnormal Boys Club (不正常俱樂部) at National Chiao Tung University in Taiwan during the 1960s (M. Lin 1996).[1] The ABCs, as they called themselves, were an informal group of eight male graduate students in the Institute of Electronics (電子研究所). The ABCs were abnormal because none had left for the United States after completing their undergraduate degrees at National Taiwan University. At that university, commonly called Taida, students were well aware of the oft-repeated rhyme: "Come! Come! Come! Come to Taida! Go! Go! Go! Go to America!" (Hsieh 2017, 27). The ABCs also stood out because the group included a mix of mainlanders (*Waishengren*) and islanders (*Benshengren*) (Fig. 1.1).[2]

The terms "islander" and "mainlander" had gained currency in Taiwan during the two previous decades (Yang and Chang 2010). The families of islander students had lived in Taiwan for as long as three and a half centuries and, together with indigenous peoples, had endured the Japanese colonial period (1895–1945) when they were all judged to be subordinate (Lo 2002; P. Chen 2006).

Mainlander families began moving to the island in 1945 when the Republic of China (ROC), led by the Kuomintang (Nationalist) Party (KMT), formally gained control of Taiwan and its surrounding islands and named them a province of China. Then in 1948–1949, when the government relocated itself to the new province until it could retake the

[1] The English name was coined by one of the members. Over time, it shortened to "Abnormal Club." Wang Huan-San Samuel, interviewed by Shih Po-Jen Bono, October 11, 2022. See also HC iTaiwan Forum (2019).

[2] The islanders were Chew Tsai-Hsin (邱再興; 1940–), Wang Hwan-San Samuel (王晃三, 1941–), Chen Long-Ing (陳龍英, 1941–2020), Chiang Chao-Yen (江昭彥, 1941–2021) and Chen Tseng-Hsin (陳增興, 1941–?). The mainlanders were Hu Ding-Hua (胡定華; 1943–2019), Hsieh Ching-Chun (謝清俊; 1941–), and Lin Chao-Wu (林朝武, 1940–?). Information from M. Lin (1996) and Chew Tsai-Hsin, interviewed by Chang Kuo-Hui, January 28, 2021. Chew mentioned that there were a few more ABC members, but additional names have not appeared in reports, news, books, or magazines.

K.-H. Chang et al., *Engineers and the Two Taiwans*, Synthesis Lectures on Global Engineering, https://doi.org/10.1007/978-3-031-59766-4_1

Fig. 1.1 Abnormal Club, National Chiao Tung University (Courtesy of Hsieh Ching-Chun)

mainland from the People's Republic of China (PRC), more than a million mainlanders fled across the Taiwan Strait, amounting to roughly 13% of the combined populations (Yap 2021; Yang and Chang 2010). Judging that island provincials were not prepared to implement Chiang Kai-shek's program of "national reconstruction," the KMT government generally assigned mainlanders to leadership positions. A 1945 report asserted that the Japanese colonialists had limited the local population to low-level education, using vocational schools as "instruments to enslave compatriots" (Ou 2002, 7–8). The "low-level education," according to a policy maker, had prohibited an overwhelming number from gaining opportunities to attend middle schools and prepare for higher education (Hsueh 1990, 300).

Hu Ding-Hua (胡定華; 1943–2019) was a mainlander ABC born in Nanking, the son of a lieutenant general in the air force. The Hu family relocated to Taiwan late in 1948 as part of a group of several thousand air force families (S. Wu 2019, 35). Hu successfully gained admission to Taida in 1959 and then to Chiao Tung for his master's degree. Although he traveled to the United States afterwards to complete a Ph.D. in electrical engineering at the University of Missouri, he soon returned to accept a faculty position at Chiao Tung. He had decided to return in 1970 after following the news about the Kent

Fig. 1.2 Hu Ding-Hua in 1971 & 1978 (Courtesy of Pan Wen-Yuan Foundation and The Central News Agency, Taiwan)

State shootings in Ohio, where U.S. National Guardsmen had fired on protesting students, killing four (S. Wu 2019, 61; Hu 2011) (Fig. 1.2).

Chew Tsai-Hsin (邱再興; 1940–) was one of the islander ABCs, born to a working-class family in Keelung.[3] His early childhood memories included hiding in air raid shelters during U.S. bombing attacks. During the early post-war years, his parents sold produce in wet markets, his two elder brothers were dock workers at Keelung Harbor, and he himself peddled popsicles and snacks (Chiu 2015, 37–39, 44–45). Impressed with his performance in elementary school, one of his teachers suggested that he might be able to pass the entrance exam offered by the top high school in Taipei—Jianguo (建國中學). Chew took and passed the test, gaining admission.

At Jianguo, Chew identified a career pathway for himself that differed from both his mainlander classmates and the fewer islander boys who were present. Mainlander families were often preparing their sons to attend elite universities and then go on to graduate schools in the United States (Chang 1992). By the mid-1960s, for example, at least 75% of students who studied engineering at Taida did indeed then go abroad after graduation.[4] By that time, those mainlanders who went to the United States often established footholds

[3] Chew Tsai-Hsin, interviewed by Chang Kuo-Hui, January 28, 2021. Chew indicated that he prefers the spelling Chew to Chiu, which has appeared in publications.

[4] Chi Guo-Chung and Lin Mou-Shiung, interviewed by Chang Kuo-Hui, March 31, 2018 and November 22, 2023. We also draw on two accounts to estimate that during the 1960s roughly 75–90% of undergraduate NTU students in electrical engineering (EE) went abroad after graduation in pursuit of advanced degrees (C. Wang 1997; S. Wu 2019).

Fig. 1.3 Chew Tsai-Hsin in 1973 & 1996 (Courtesy of Chew Tsai-Hsin and the Alumni Voice of Chiao Tung University)

for their families to follow, typically families that did not identify with life on the island or envision futures for themselves there (Chang 1992).[5]

The families of islander boys who gained admission to Jianguo tended to be wealthy local elites. They were often positioning their sons to seek local degrees in medicine and respected local careers as doctors (S. Wu 2019, 44; Lo 2002). As both an islander and not one of the elites, Chew identified a pathway that might lead to upward class and career mobility—through electrical engineering (Chiu 2015, 64–65). In the 1960s, for any islander to gain admission to a national university, let alone Taida, was still an exceptional achievement. Most who sought education beyond secondary school found their access limited to junior colleges and a few provincial colleges. Chew's performance at Jianguo warranted his admission to Taida, but his identity as an islander and his socio-economic status still made him stand out. He worked his way through the university as a teacher in a cram school and then stayed home again for graduate school, winning admission to Chiao Tung (Fig. 1.3).[6]

While Hu and Chew were gaining engineering credentials at Taida and Chiao Tung, the KMT government was focusing new attention on industrial development in the island province. Its interest was prompted by advice and pressure from the U.S. government to develop a market economy and accelerated by an expected 1965 announcement that the U.S. government would be ceasing foreign aid to the Republic of China (T. Wu 2023, 290–291, 350–351). The United States had begun turning its strategic attention to the PRC, building a new relationship that would lead to the ROC losing its seat in the United

[5] Chi Guo-Chung, Roger Liao and Lin Mou-Shiung, interviewed by Chang Kuo-Hui, April 10, 2018, April 3, 2018 and November 22, 2023.

[6] Chew Tsai-Hsin, interviewed by Chang Kuo-Hui, January 28, 2021.

Nations in 1971. Eight years later, the U.S. government would formally stop recognizing the ROC as the government of China.

While it remained in control on the mainland during the 1930s and 1940s, the KMT government had pursued national reconstruction most prominently through state ownership of large production infrastructures (Kirby 2010). After the war, it extended this practice to the island province, taking over nearly 500 private Japanese companies and merging them together into companies owned by national, provincial, and county governments, as well as by the Kuomintang Party itself (Hong 2016a). And between 1950 and 1965, it spent the bulk of nearly $1.5 billion USD in aid from the United States on large state-controlled projects for the province in transportation and communication (Jacoby 1967). Islanders participated in these projects as they had attempted to participate in Japanese companies, climbing infrastructures from low-level positions and struggling with glass ceilings (C. Lin 2012, 114–119; Kao 2016, 166–179).

The KMT government had not prioritized the development of manufacturing and service industries in what it considered to be a temporary base of operations. Instead, it had recently begun to build a private industrial sector by inviting U.S., Japanese, and European companies to come build factories, promising them low wages, virtually no environmental regulations, and favorable trade policies (Lin and Chen 2021, 38–40; T. Wu 2023). While Japanese companies were required to invest in joint ventures with local companies, U.S. companies had enormous freedom to come and go as they pleased (Amsden and Chu 2003; S. Liu 2012 [1992], 371; Hong 2016b). However, with the anticipated loss of U.S. aid and support prompting concern about the balance of payments and interest in increasing exports, the government gave special attention to new industries, especially electronics.

Hu Ding-Hua and Chew Tsai-Hsin both contributed to the development of an electronics industry but via sharply distinct trajectories. Hu the mainlander found himself participating in, and sometimes leading, government initiatives to support the development of large state-owned organizations. He focused on producing competent “talents” (人才).

Meanwhile, Chew the islander plotted a pathway into the production of electronics products that was confined wholly to smaller, mostly family-owned companies beyond the interest or reach of government support. He did start out as an engineer at a consumer electronics plant built by a U.S. semiconductor company, but despite earning a salary that was far greater than those of his professors at Chiao Tung, he did not want a career in a foreign-owned company. As a child, he had witnessed U.S. soldiers harassing local women on the street, and was outraged as a high school student when a U.S. military court in 1957 acquitted a soldier who had shot and killed a local man (H. Lin 2021, 66–72; Chiu 2015, 74–75, 91–92). Later remembering a demonstration in which he had “happily watched the American flag burned and American cars destroyed,” he explained that “we [locals] were no longer bullied by the American and knew how to resist” (Chiu 2015, 75). After three years at the foreign firm, Chew left for a small local company founded by another graduate of Taida in electrical engineering, accepting a 50% pay-cut in the process.

Over the next two decades, technological capabilities and domestic production in electronics on the island territory would expand rapidly enough to attract international attention and warrant claims by supporters that an industrial "miracle" had taken place (Saxenian 2006, 24, 126). As business analysts debate the case, Taiwan was able to make itself a center for the electronics industry for such reasons as its development required fewer natural resources than chemical and other manufacturing industries and it relied to a greater extent upon educated workers and advanced production techniques and devices that the island was able to provide (Cheng 1994, 168; Amsden and Chu 2003, 16–17, 102; Norberg 1999, 209; L. Wang 2013). Certainly, between the mid-1960s and the end of the 1980s, gross national income increased by more than a factor of 15, with international trade increasing by more than a factor of 50 and with electronics as the iconic leader (DGBAS 2023). Yet the emergence of electronics in Taiwan also included another crucial dimension that we explore in this book, the inclusion and rise of islander engineers.

The distinct trajectories pursued by Hu Ding-Hua and Chew Tsai-Hsin as engineers would prove to be a distinction with a significant difference for the identity of Taiwan. The separation between large organizations and companies with government support and smaller ones without it would become a space for a new type of relationship between mainlanders and islanders to develop. In a sense, the "abnormal club" that included islander engineers alongside mainlander engineers would get much bigger, challenging the prior hierarchy and providing a model of juxtaposition for Taiwan as a whole. Yet juxtaposition was not full integration, the elimination of hierarchy. Through the material work of building novel infrastructures for a unique industry, electronics engineers from the 1960s to the 1980s would play key roles in animating not just one Taiwan, but two.

An Experiment in Attaching Learning From STS

This project began as a straightforward inquiry into the emergence of engineers on the island of Taiwan. We were guided by an overarching question: how do engineers and Taiwan relate to one other? This question posed issues in both directions, especially given that the island itself has experienced multiple and contested geographical identities over the past two centuries. Looking in one direction, might following episodes in the emergence of engineers on that island tell us something about its changing positioning in relation to other territories, including eventual emergence as a thing, or things, unto itself? And looking back in the other direction, might the contents and timing of emergent infrastructures of engineering education and work on the island tell us something about the agencies of engineers as technical practitioners, not only within Taiwan but also possibly beyond? That is, to the extent that developments involving engineers overlapped and intersected with developments involving the island, we could query how the technical work of making engineers and doing engineering in Taiwan may have also pointed engineers in particular directions, attaching material and value commitments in the process.

As we will see, the evolving geographical positioning of the island in relation to the Chinese mainland, European countries, Japan, Korea, Southeast Asia, and the United States, all of which were also evolving simultaneously, raised numerous and continuing questions of power and politics. It might seem obvious that the long process of materializing Taiwan—making it real as Taiwan—has been thoroughly political in content, both externally and internally. Accordingly, any analysis of that process must attend to politics, and most accounts do.

As such, inquiring into the positioning and contributions of engineers might appear to make little sense. Engineers do not do politics; they avoid it with a passion. Or so they have tended to claim as the figure of the engineer emerged and spread through European and EuroAmerican contexts and initiatives. Engineers comfortably, and regularly, assert that they do technical work only. The continuing force of this stance can be seen in the tendency for engineers to treat only deviations from it as requiring analysis and explanation, e.g., via technical mistakes, ethical failures, and so on.

Yet notice that the claim to do technical work only actually introduces a significant and revealing political distinction. It parses out a particular kind of politics. When engineers say they do not do politics, they are typically referring to partisan politics, participation in political parties. Partisanship necessarily presumes disagreement and conflict. By construing engineering work as exclusive of politics, engineers are resisting identification with actions that might position them in conflict with others. As they present themselves and their work, engineers do not have opponents. They do not do conflict. They do, however, do commitment, passionate commitment. Engineers affirm their resistance to conflict by passionately asserting that their work produces benefits for all, unless they make mistakes or someone somewhere perverts it. Put otherwise, through a kind of normative holism, engineering work is omnibenevolent, helping everyone through an ethical commitment to material provisioning (Downey 2012; Kleine 2020; Smith 2021). Engineers provide the material world with the things its inhabitants need and want.[7]

Material provisioning for all includes an assumption of shared land. Land is not singular, however, but plural—lands.[8] It becomes possible to ask if, in the process of material provisioning for this land and its people versus that land and its people, might engineers be positioning themselves and their practices in ways that contrast from one another

[7] The sense of holism carries into subfields of engineering work. The mechanical engineer does not work against the civil engineer. Rather different fields of engineering simply divide up approaches to material provisioning for omnibenevolence.

[8] The notion of lands here includes waters, including those waters that separate lands. Also, physical territory is the default attachment, but the specificities in its meanings are to be discovered rather than defined in advance.

geographically? There are of course a multitude of ways in which engineers and engineering practices might and do contrast with one another. But here, in this project, we focus attention on the attachment of engineers to specific lands, calling these attachments "geographical identities."[9]

A commitment to one's land is a passionate commitment. As scholars in the field of science and technology studies (STS), we are predisposed as authors to see a specific attachment to a specific land as political in content because it enacts differential power dynamics in relation to lands and peoples. This project takes a different turn. Recognizing that engineers tend to construe politics as partisan in content, it starts by tracing and following the specific attachments to a given land, exclusive of others, questioning linkages among knowledges, identities, and commitments. In other words, we explore attachments between engineers and lands without defining politics in advance.

This approach has two benefits. First, it allows us to focus on potential "congruences" (Faulkner 2009) between the technical identities of engineers as practitioners and their geographical identities as inhabitants of a particular land. That is, through the persons of engineers, engineering practices gain contents that go beyond technical dimensions, narrowly construed. One can see how the work of engineers does have nontechnical dimensions that carry enormous force and merit analysis.

The first benefit feeds the second, helping to position this book for engineering audiences. This project is an experiment in adding or attaching learning from STS to how engineers currently understand their knowledge and identities as engineers. It seeks to warrant and stimulate critical reflection by engineers on their knowledge, identities, and commitments. As such it enacts critical participation, an approach to the scholarship of STS making & doing that involves inflecting dominant knowledge forms and practices through expressions and flows of STS learning (Downey 2008; Lucena et al. 2010; York 2018; Wisnioski et al. 2019; Smith 2021; Downey 2021; Shih 2021). Might coming to recognize that congruences between geographical identities and technical identities played a role in the emergence and career trajectories of engineers in Taiwan bring awareness to how other powerful identities have similar issues of congruence as well, e.g., identities based on gender, race, sexuality, ethnicity, field or discipline, site of education, positions within organizations, colonial/postcolonial relations, global North/South differences, and so on?[10]

[9] With the term "identities," we are referring to relations among entities (id+entities). The term "geographical identities" refers to relations among entities that are spatial, usually terrestrial, in content.

[10] Note that the islander engineers and mainlander engineers remained predominantly male during the period considered here. We hypothesize that reflecting critically on geographical identities could help open engineers to critical reflection on gender identities.

Opening Up Infrastructures of Engineering Formation and Work

Our central method has been to trace the development of infrastructures of engineering formation (including education and training) and engineering work across Taiwan, looking for key episodes in the changing relationships between emergent engineers and the island. Infrastructures of engineering formation include physical, organizational, and abstract structures ranging from school buildings to government ministries to course curricula and contents. Infrastructures of engineering work range from actual work practices to organizational roles on the job and in professional societies to the positioning of engineering work in relation to larger missions.

We explore five formative episodes: the Japanese colonial period, early KMT period, development of the electronics industry, emergence of professional societies in the United States, and, finally, a period characterized by the flows of engineers from Taiwan to the mainland. Infrastructures of formation and work in these episodes enabled particular images of engineers and engineering to emerge, travel, and become dominant for those who faced them. Also, prospective and working engineers who passed through such infrastructures or otherwise encountered them were challenged to accept forms of knowledge that pointed their work and careers in particular directions, with associated identities and commitments.

The episode in each chapter inquires into how prospective and working engineers actually did respond to the evolving challenges and expectations they encountered, navigating both technical and geographical commitments. Through a selective collection and marshalling of relevant evidence, the chapters show together how congruences between technical and geographical identities had the effect of attaching emergent engineers to the land of Taiwan as it evolved to become a thing, an entity, unto itself. The level of detail in the episodes might suggest that they are histories. They are not. They are sharply delimited investigations of relationships between the pathways of engineers into and through engineering and their connections to the land of Taiwan.

The episodes in Chaps. 2 and 3 explore the struggles of would-be local engineers and engineering technicians to climb industrial infrastructures that were controlled by others. During the 50-year colonial period examined in Chap. 2, the Japanese empire understood its engineers as technical talents who had knowledge of machinery, and engineers on the island nearly always came from the Japanese mainland. Most local boys and men that went to work in machine-based industry were educated only in common schools and had to join as manual laborers. For the most part, those who were able to gain access to some machine knowledge in new industrial schools had to work as assistants to Japanese technicians or, later, travel to Manchukuo.

Yet some were able to avoid direct Japanese supervision and become "freelance" workers, akin to doctors, who established small companies that hired local labor and built spaces to operate through contractual relationships with Japanese companies. And some freelancers who worked successfully outside of Japanese companies began to embrace the land of Taiwan apart from the empire by claiming the identity of professionals, a category of person that is typically seen as detached from geography. As Taiwan became a base for southern expansion and key contributor to the Greater East Asia War, many local companies flourished. And by 1945, local industrial technicians and would-be engineers were contributing significantly not only alongside Japanese companies but also within them.

The Republic of China differed from Japan by initially positioning its engineers as elite knowers for national reconstruction whose commitments to industrial productivity were realizing science as moral enterprise. Examined in Chap. 3, when the Kuomintang government gained control of Taiwan as a new province after the war, it quickly took ownership of private Japanese companies and consolidated them into large state-owned companies run by elite engineering managers who were credentialed with university degrees. And when the government fled to Taiwan in 1948–1949, the many engineers who went along transferred mostly to the provincial government, the army, and large companies that received from the government monopoly control over their industrial arenas. Local company owners on the island eagerly established organizations to demonstrate their own commitment to science and desire to help resume industrial development. However, arriving officials judged that the low-level education they had received from the Japanese prevented them from becoming acceptable knowers for national reconstruction.

The provincial government reframed education around a hierarchy of science learning, treating vocational learning as applied science that effectively privileged mainlanders over islanders. Introducing science to children at the primary level, it repositioned industrial schools as vocational schools, elevated some to junior colleges and provincial schools, and made the imperial university a national university. Wholly taking for granted that mainlander engineers would remain in control over industrial arenas by virtue of their higher education and control of large companies, the government permitted economic initiatives by local companies and the formation of local engineering technicians and engineers. Local engineers and would-be engineers turned their sights to small companies, once again forced to focus on climbing industrial infrastructures from the underside,

The episodes in Chaps. 4 and 5 follow the elevation of islander engineers alongside mainlander engineers as both participated in the unique industrial organization of electronics technologies. The initial development of production infrastructures for the electronics industry was led prominently by mainlander engineers. Chap. 4 shows how they played principal roles in establishing National Chiao Tung University, the Modern Engineering and Technology Seminar, the Industrial Technology Research Institute and its Electronics Industry Research Center, the Hsinchu Science Park, the Ten Major Construction Projects, and the spinoff of United Microelectronics and other state-supported enterprises. In contrast with production models in Japan and Korea, the KMT vision for Taiwan planned for the centralized production of semiconductors by large companies and the distributed production of products by small companies whose economic contributions would not raise political concerns during the period of martial law. A rapid expansion in junior colleges followed, particularly benefitting islanders. Still, the given hierarchy between state-sponsored companies controlled by mainlanders and small private companies operated by islanders made expanding technical education for islanders an activity that would affirm the existing order, not threaten it.

But the electronics industry was not like the factory-based industries of the past. Growing quickly with a distribution of expanding openings and new niches, the functioning of nodes depended on the filling of openings, and small companies multiplied rapidly as those relationships mutated. The emergence of a transnational identity for Taiwan as an electronics powerhouse made it into a thing on the world stage. For mainlander engineers, that identity could be linked to that of the Republic of China. For islander engineers, pursuing work and careers in electronics contributed to the emergence of something different than another version of China led by mainlanders, minimally an image of Taiwan as a thing unto itself. Small and medium-sized companies became essential partners alongside mainlanders in an industry that effectively animated two Taiwans, making the "abnormal club" of engineers who worked for Taiwan a lot bigger.

While islander engineers living and working in Taiwan were not permitted to announce their differences from mainlanders during the period of martial law, Chap. 5 shows how engineers who relocated to the United States for advanced education and work were able to do so. While embracing the identity of professionals once again established a safe distance from partisan politics, this time the naming of professional organizations introduced geographical identity as a source of difference. Mainland engineers traveled first, often taking with them families that did not identify with or want to live in the remote, rural province that they considered to be temporary refuge. They tended to join professional organizations that explicitly linked themselves to China, e.g., the Chinese Institute of Engineers. As the emerging electronics industry and waning of the KMT dream to retake the mainland were making Taiwan a distinct entity in the world, some formed the Monte Jade Association. It identified not with Taiwan but with its tallest mountain, meaning that engineers could lead the world's Chinese diaspora from Taiwan without claiming to be "of" Taiwan.

Meanwhile, some early islander engineers who went to the U.S. did choose to embrace politics in partisan terms, joining organizations that identified with Formosa and, sometimes, advocating for independence. Later, as Taiwan became a thing, they embraced the land by naming organizations "Taiwanese," e.g., the North American Taiwanese Engineers' Association. While mainland engineers were not excluded, they were increasingly identifiable as "mainland Taiwanese." Members of Taiwanese organizations tended to be islander engineers who identified increasingly as "native Taiwanese" or, when in Taiwan, as "local Taiwanese." Professional organizations in the United States thus offered a way for islander engineers to elevate their visibility and claim their identities as such just as they were becoming essential contributors to an emergent industry within Taiwan. At the same time, for many engineers, likely a majority for both mainland and native Taiwanese, joining a named society meant embracing the kind of politics that engineers tend to resist.

These engineers chose low profiles that let their technical work alone express their material and value commitments.

During the 1990s, the distinction between mainlander engineers and local engineers appeared to blur as both participated enthusiastically in extending Taiwan-owned companies with Taiwan-developed technologies across the Taiwan Strait to mainland China. Examined in Chap. 6, enrollments in engineering degree programs at Taiwanese universities basically exploded as the KMT government elevated junior colleges to the status of universities. A preponderance of mostly local students pursued degree programs in electronics engineering and computer science. Also, over the next decade large numbers of native Taiwanese working abroad returned, finally overcoming the sense that career opportunities and family responsibilities had to pull one in opposite directions.

By establishing and maintaining a physical presence on the mainland, mainland Taiwanese engineers and local Taiwanese engineers carried with them distinct images of Taiwan's relationship with China. They effectively traveled across the strait on two pathways. One pathway sought to maintain centeredness and control in Taiwan while the other sought to build productive power with mainland China. The manufacture of semiconductors became a key site of tension between them. Then engineers from the mainland itself enacted yet another geographical identity as the Chinese government increasingly asserted its own vision of the meanings and implications of cross-strait relations.

Taken together, the five formative episodes demonstrate that congruences between the technical identities and geographical identities of engineers have both granted them privileges and assigned them responsibilities as they helped to make and re-make the land of Taiwan. Isolating the relationships that emerged between engineers and the island through infrastructures of education/training and work, Chap. 7 shows how those privileges and responsibilities extend through the present.

A Tension on the Frontier

To recognize the arrival of Japanese infrastructures of formation and work in Chap. 2 as novel and forceful, it is important to call attention to the infrastructures that were already in place or emergent or otherwise at stake. In particular, when the Japanese military arrived in 1895 to take control, they may not have known how close the Qing Chinese province had come in recent years to producing its own engineers. The experiences of that initiative made visible a longstanding tension between the official dominance of classical learning and the variable opportunities that came with the remoteness of the province itself.

Eight years earlier, a governor appointed by the Qing court had ordered the construction of a Hall of Western Learning that included daily studies of "calculation, surveying, and manufacturing" alongside English, French, geography, and history (Kuo 1996; Hsu 1993). Governor Liu Ming-Chuan (劉銘傳 1836–1896) was intent on building a military defense for the island as a whole. He was leading an initiative on behalf of the court that had started two decades earlier when the Japanese navy landed in a remote area to punish an indigenous group that had killed 31 shipwrecked seamen. The Qing initiative then gained urgency in 1884 when the French navy invaded the north expecting little resistance, planning to include the island in its efforts to expand colonial control across what would become Indochina. A career soldier, Liu was dispatched to command the defense.

Picturing the island as a whole was one outcome of a "self-strengthening" movement on the mainland (1865–1895). Its advocates sought to reimagine the Qing empire as a singular whole with definable borders and, hence, the prospect of greater "internal" integration and control from the center. Among other things and amidst considerable resistance, getting there included building arsenals and shipyards as well as telegraph and railroad networks (Yue 1999; Kennedy 1974; Kuo 1996). An advocate himself, Liu had earlier "become convinced of the clear superiority of Western weaponry and technology" while leading armies on the mainland (Chu 1963, 37). During the successful eight-month war with the French around Keelung, his soldiers rebuilt the forts guarding that city and what would become Tamsui in the north as well as Takou and what would become Kaohsiung in the south, outfitting all with new Krupp and Armstrong cannons. He also established an arsenal in Taipei to service these cannons and produce small arms and ammunition (Chu 1963, 38).

After the war, as newly-appointed governor, Liu went to work integrating the island as a strategic military resource, a buffer for the mainland. Portions of the island along with surrounding islands had previously been a subordinate prefecture of Fujian province across the Taiwan Strait. During the war, Liu was made governor of that province and then afterwards reassigned to a separate office of governor of "Taiwan" until it became a new province in 1887.

Leading projects to bolster defenses and integrate the province put Liu into regular contact with at least seven European engineers and a Chinese-American engineer. Greatly interested in their work, he personally accompanied a German engineer on the first four miles of survey work for a railroad project (Chu 1963, 49). The new Hall of Western Learning took a step in the direction of engineering education by training technical personnel to work for the arsenal, railroad, mines, etc. (M. Liu 1958, 297).

At the same time, foreign engineers on the island found themselves occupying uncertain positions. While they were key contributors to projects contracted with the province, they could not direct the technical work. They could not implement strategic design and production decisions nor supervise the soldiers who often provided labor. They had to accept subordination to officials with different knowledge. Their ambiguous status illustrated the difficulties Liu and other advocates for self-strengthening in the province faced in attempting to build new infrastructures of formation and work around Western learning. Foreign engineers appeared as both a key point of contact and a key indicator of contrast.

At that time, the dominant formal infrastructures for learning and work across the province offered classical education and literary training. Practices of classical learning aimed at reading and memorizing texts from the Four Books and Five Classics of neo-Confucian philosophy. Such learning was made available to boys of means in academies that could be as small as a single room in a teacher's dwelling. By building the capability to read and interpret up to 4,000 characters in the ancient texts, participants learned one's "way" as a journey that was both spiritual in content and encased in a material body. It entailed engaging matter through a range of social and ethical practices. Emulating ancient manifestations of virtue, closely-guided learning stressed the investigation of things in order to make thoughts sincere, hearts rectified, persons cultivated, families regulated, states rightly governed, and a world imagined at peace (McCausland 2011).

Accomplished reading and recitation could qualify one to sit for municipal and provincial exams that in turn could certify one's elevated status as a gentry scholar prepared for community service and leadership. Highly accomplished understanding and interpretation of the neo-Confucian texts could qualify one to sit for imperial exams that could lead to a court-appointed official post and working career as a scholar-official. Overall, many years of study produced gradual advancement in both degrees and official positions and, in principle, attachment to the empire. Ideal leaders were thus literati who were expected to embody illustrious virtue and renovate the people by extending knowledge to the utmost, enabling advancement to the supreme good (Lamley 1964, 375). In other words, in the longstanding hierarchy of occupational categories applied to commoners, those who became literate could qualify as gentry (*shi*). As such, they were clearly superior to those named by three categories of nonliterate workers: peasants (*nong*), artisans (*gong*), and merchants (*sang*) (Lamley 1964, 15).

However, the land of Taiwan was not the mainland. It had developed as a remote frontier. Its physical separation from the mainland across difficult waters had marked its inhabitants for centuries as beyond the boundaries of dynastic empires. Officials in the Ming dynasty court (1368–1644), for example, did not have a specific name for the island for most of its reign. Early on, they often referred to it generally as part of the "Foreign East" before fixing on the name Taiwan for land it claimed (Chou 2012, 151, 162–170; 2007, 93–148). After all, outside military threats to dynastic integrity had long tended to come overland, from the west and north.

In addition, the Ming court did not need its empire to expand in order to maintain its legitimacy. Establishing, maintaining, and challenging imperial governance in neo-Confucian terms depended more upon the demonstrated superiority, behavior, and privileges of the governing center rather than on the acquisition and control of new territories on the periphery. A heavenly mandate made China the center of human civilization, the so-called "Middle Kingdom" between heaven and lesser peoples. So while the Ming court certainly maintained a visible military presence within the empire, it did not have to conquer and occupy the lands of distant barbarians in order to demonstrate its worth to itself.

With marginality came both flexibilities and irregularities. Court policies and practices had the effect of making harbors on the island frequent destinations for the desperate and the unusually entrepreneurial. In its early years, the Ming court had built a substantial fleet of ocean-going warships but found the initiative to be too expensive to maintain. During the 1400s and 1500s, it actually turned away from the eastern waters, banning sea travel, making ship-building a capital offense, ordering the destruction of ocean-going vessels, and arresting any who used them. Many traders and pirates observed the sea ban in its violation, populating and using the eastern waters, including small portions of the big island territory (K. Chen 2005, 68–69; Andrade 2008, 75, 224). Over time, more than 1,000 Hakka Han Chinese peasants migrated from eastern Guangdong province, attracted by the vast stretches of level, arable plain that far exceeded their source locations (Lamley 1964, 26). They acquired land from indigenous peoples through force and barter to cultivate and trade rice and sugar.

The land of Taiwan first merited formal attention from the Ming court when the Dutch East India Company arrived in 1624. Its warships carried a Western expansionist image of empire built upon practices of settler colonialism. The company was competing with Portuguese and Spanish warships and administrative networks for the control of territory and acquisition of wealth and power through trade. Threatening military intrusion on the mainland, the company gained the court's approval to establish a colony on the southern portion of the island. Following Portuguese navigators and cartographers from the previous century, it called the island "Formosa," or variations of that name (Ang 2006, 73, 74–76) (Fig. 1.4).

The company made its colony a frontier center for trade. The southern port became a successful collection and transfer point for traded or smuggled Chinese porcelain, as many as 200,000 pieces per year (K. Chen 2005, 390). The company collaborated with Chinese and Japanese smugglers to acquire silk and silver and send deer skins to Japan to make scabbards and other products. And it took decisive steps to expand the agricultural production and export of sugar and rice within and beyond the colony. It did this by pushing Hakka peasants off their land and toward the mountains. It then collaborated with the Ming court to recruit impoverished Hoklo Han Chinese peasants and artisans from eastern Fujian province and from Batavia, the center of Dutch colonial operations in the region (Kao 2016, 27–28; K. Chen 2005, 383–384). As many as 100,000 Hoklo Chinese, nearly all men, came to work the fields and production facilities. Colonial officials deemed them to be the "only bees on Formosa that [gave] honey," meaning that they participated largely without resistance (Andrade 2008, 75, 224, 195).

Had the Dutch outpost turned into long-term settler colonialism, it might have prioritized what was becoming known on the mainland as *Rangaku*, or Dutch learning. Certainly, missionaries established schools to add to the local and imported peoples both Christianity and enough of the Dutch language to facilitate trade. But the company made no sustained attempt to construct and promote educational infrastructures to introduce European knowledge about, especially, medicine and science. It did protect trading activities by building a massive stone fort in Anping and it competed for sixteen years with the Spanish, who had established a small trading settlement in the north, before successfully

Fig. 1.4 Fort Zeelandia, constructed by the Dutch East India Company 1624–1629 (Courtesy of National Museum of Taiwan History)

driving them out. While the company built roads in the western corridor to facilitate trade, it maintained a geographical vision of the land as built around the colonial center in the south. To that end, it avoided interacting with the roughly 100,000 indigenous peoples who populated and actively resisted intrusions into the central mountains and much of the narrower, steeper eastern corridor (Andrade 2008).

In 1661, a Ming scholar-official turned military leader who was fleeing Manchu invaders took refuge on the island and laid siege to the main fort in Anping. After a nine-month battle, the forces of Cheng Cheng-Kung (鄭成功, 1624–1662) successfully ejected the East India Company from southern Taiwan and then over the next several years dismantled Dutch governance that had extended north. Known by the Europeans as Koxinga, Cheng became a much-debated figure for the identity of the land of Taiwan. Fifteen years prior to the victory in Anping, the Ming court had lost control of its empire to Manchu invaders, who then claimed the mandate of heaven for a new Qing dynasty. Given that change, ridding the island of Dutch colonialists could have been construed as founding an independent Taiwan. However, it could also be counted as one of the last acts of Ming resistance, repossessing the land for imperial China prior to yielding completely to the Qing court in 1683. The military victory could even be seen as staking an early claim for Japanese rule on the island. Cheng Cheng-Kung was born half-Japanese, and the island's geographical positioning could construe it as the southernmost member of an extended archipelago. To what frontier land Koxinga's forces were attached was a question that never got a settled answer (Fig. 1.5).

In contrast with passive Ming interest, in 1684 the Qing court formally incorporated occupied areas of the island as a prefect of Fujian province, managed by a prefectural capital in Tainan, in the south. Yet during two centuries of rule, Qing oversight did not become fully focused attention until after the incursion by the Japanese navy and attack by the French. Until 1887, the prefect (which then became two) included eight counties and four independent subprefectures. Fully 60% of the counties were still occupied sparsely by indigenous groups, including those that Japanese officials later called the Ami, Atayal, Bunun, Kavalan, Paiwan, Rukai, Puyuma, Saisiyat, Tao, Tsou, and Thao (Speidel 1967, 276).

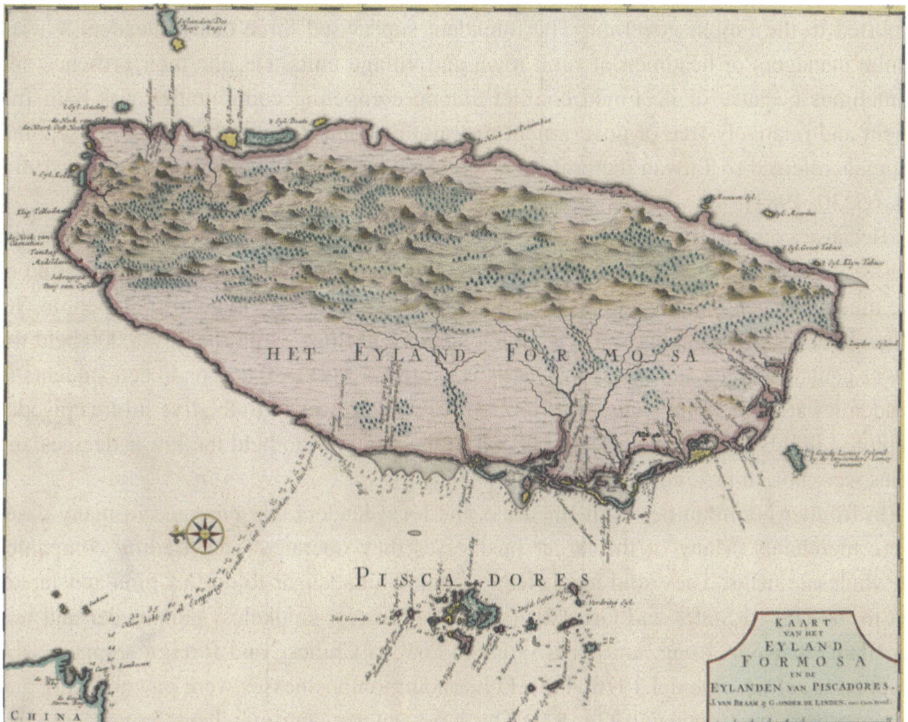

Fig. 1.5 Map of Formosa and the Pescadores, 1726 by J. van Braam and G. onder de Linden with Privilege (Courtesy of National Museum of Taiwan History)

Most Chinese inhabitants on the island frontier maintained their primary attachments to their home territories, treating their lands as adjuncts to the mainland. Chinese settlements grew in haphazard, disjointed ways. For its first 80 years, the Qing court officially restricted emigration across the strait, yet growing illegal emigration produced increasing populations. Most emigrants still came from two areas of Fujian province and from southeast Guangdong province, seeking work and income on and around good arable land. And after the court relaxed emigration in 1760, thousands of persons arrived each spring from the mainland to work temporarily in the emergent oolong tea industry, which eventually replaced sugar as the main commodity (Davidson 1903, 385; M. Lin 2008, 19–56). Emergent communities partitioned around geographical and surname identities (Ownby 1990, 91). Despite continuing attachments to the mainland, the Qing court actually treated members of those communities as overseas Chinese, meaning that they had to endure the same restrictive regulations as emigrants who had gone to southeast Asia (Lamley 1964, 33).

Court-sponsored governance on the frontier was present but loose and irregular. The chief administrative officer was the "intendant" in the south, who for most of the period reported to the Fujian governor. The intendant supervised three district leaders as well as the managers or headmen of rural town and village units. Despite their presence and sometimes because of it, armed conflict among competing communities was both frequent and relatively free of government restraint or control. And not even Qing scholar-officials referred to Taiwan the prefect as an island, a geographical whole (Lamley 1964, 64, 65, 30; 1981).

Because comparatively few gentry scholars emigrated to the prefect, academies for classical learning grew slowly and most exams took place on the mainland. By the end of the nineteenth century, the Chinese population was approaching three million people. Yet only about 350 held the highest academic degrees and titles while roughly 5,000 held the lower gentry degrees and titles, and perhaps another 25,000–30,000 had been students in academies at some point (Lamley 1964, 13). Crucial, however, to the five future episodes explored in this book, a significant portion of the gentry who held the lower degrees and titles were not, in fact, educated literati.

In frontier communities built for trade, the local leaders that emerged in many cases were merchants. Many of the larger businesses they operated were trading companies or wholesale firms. They sold local rice to south China, sugar to north China and Japan, tea to the United States and Great Britain, camphor (for smokeless gunpowder and bug repellant) to Hong Kong, and locally-mined coal to Chinese and foreign shippers who stopped to refuel (Speidel 1976, 445). Other common businesses were pawnbroking and money changing (Finegan 1978, 82). The most enterprising merchants conserved their wealth by investing business profits in lands or spreading their investments across different business, adding such products as paper, paint, and dried fruit (Finegan 1978, 83, 84).

Wealthy merchants became powerful local figures. They contributed funds for the construction of temples, town walls, and academies as well as for irrigation projects and other public undertakings (Lamley 1964, 70). While they assisted local officials in the collection of duties for trade, they also became known for evading taxes and government oversight. Those who aspired to gentry status were increasingly able to do so outside of the infrastructures of classical learning through the purchase of titles. Many certainly had some basic education, studying elementary subjects at home before learning business practices on the job. But by the mid-1800s, the Manchu court was in desperate need of funds and permitted greater numbers of wealthy subjects to enter into the lower gentry through direct purchase (Finegan 1978, 85, 86).

The frontier island became marked by unusually close ties between merchant and gentry groups (Lamley 1964, 12). The gentry merchants did not come through the infrastructure of classical formation and work, but their elevated status still depended upon it. And the low-level titles they received attached them strongly to their local communities but not necessarily to the empire as a whole.

The relative wealth and prominence of gentry merchants along with their relative distance from the Qing court made them especially useful allies for Liu Ming-Chuan's efforts to add Western learning and integrate the island. Liu appointed merchants to key positions in his administration and recruited them to help pay for his defense projects, including mining, telegraph, shipping, and railroad. For example, Lin Wei-Yuan, one of the wealthiest men in the province, was placed in charge of local militia, supervised land reclamation, and directed Liu's railroad project (Speidel 1976, 450). And Li Ch'un-Sheng received minor office titles in exchange for financial support for the railroad project (Lamley 1964, 12). So even while Liu Ming-Chuan was adding a new infrastructure of learning that pointed toward engineering, he still deferred to and mobilized the glories and resources of the Qing court to gain local support, indirectly affirming the continued force of classical learning. Indeed, his responsibilities as chief official in the province included directly supervising provincial exams.

Liu Ming-Chuan's initiatives in Western learning were also challenged by the images he had available to him to categorize foreign engineers and potentially produce domestic versions. Liu understood the practical work of foreign engineers through the lens of *gong*, artisans who worked with their hands (Han 2021). To him, the foreign versions were a special type of *gong* who occupied elevated positions akin to gentry. In his writings, he referred to foreign engineers as *gong shi* (工師), meaning masters who held artful work and skills (Shih 2021, 167; Liu 1958, 26, 274, 357, 359, 361–362). But the term for him was not settled. He also referred to the engineers as *yang ren* (洋人), meaning ocean men—they were like other foreign men who had traveled across the ocean (M. Liu 1958, 34, 356, 365–366). And he used the term *yang shi* (洋師), as they were masters who came from the ocean (M. Liu 1958, 34–35, 351–352, 356, 367).[11]

[11] During Liu Ming-Chuan's rule, a Chinese lexicographer, Kwong Ki Chiu (鄺其照) compiled the first English-Chinese dictionary on the mainland. *An English and Chinese Dictionary*, published in 1887, describes the "engineer" literally as "a person who manages steam engine" and "civil engineer" as "a person who draws a house diagram" (Kwong 1887). Later, the most common translation

The total of 64 graduates from the Hall of Western Learning received certificates indicating that they had indeed added new learning that enabled at least some artful work and skills (S. Wang 2004; Kuo 1996; Han 2021). But Liu Ming-Chuan did not have enough time to produce *gong shi* for the province, even if he actually did have that outcome in mind. When he stepped down in 1891, ostensibly for health reasons, his successor soon closed the Hall of Western Learning. In what was still a partitioned territory on the periphery of the empire, no alternative infrastructures were in place or emerging to challenge the authority and power still afforded by degrees and titles earned through classical learning.

Then the Japanese empire arrived.

References

Amsden, Alice H., and Wan-wen Chu. 2003. *Beyond Late Development: Taiwan's Upgrating Policies*. Cambridge, MA: MIT Press.

Andrade, Tonio. 2008. *How Taiwan Became Chinese: Dutch, Spanish, and Han Colonization in the Seventeenth Century*. New York: Columbia University Press.

Ang, Kaim 翁佳音. 2006. "Pu tao ya ren yu Formosa: Bing lun 1582-nian de chuan nan 葡萄牙人與「福爾摩沙」：並論1582年的船難 (The Portuguese and "Formosa" and the Shipwreck of 1582)." *Li shi yue kan* 歷史月刊 *(Historical Monthly)* 220: 72–79.

Chang, Shirley L. 1992. "Causes of Brain Drain and Solutions: The Taiwan Experience." *Studies in Comparative International Development* 27: 27–43.

Chen, Kuo-Tung 陳國棟. 2005. *Tai wan de shan hai jing yan* 臺灣的山海經驗 *(Taiwan's Experience of Mountains and Seas)*. Taipei: Yuan-Liou (遠流).

Chen, Pei-feng 陳培豐. 2006. *"Dōku" no dōshō imu: Ri zhi shi qi tai wan de yu yan zheng ce, jin dai hua yu ren tong* "同化"之同床異夢: 日治時期臺灣的語言政策、近代化與認同 *("Assimilation" in the Same Bed with Different Dreams)*. Taipei: Rye Field Publishing (麥田).

Cheng, Chu-yuan. 1994. "Taiwan's Economic Development Under Chiang Ching-kuo." *The American Asian Review* 12(1): 164–198.

Chiu, Tsai-Hsing 邱再興. 2015. *She de: Dian zi ye xian qu qiu zai xing de shi ye yu zhi ye* 捨得: 電子業先驅邱再興的事業與志業 *(Willing to: Career and Aspirations of Electronics Pioneer Chew Tsai-Hsing)*. Taipei: Booklife (圓神).

Chou, Wan-yao 周婉窈. 2007. "Shan zai yaobo bilang zhong: Zong lun ming ren de tai wan ren shi 山在瑤波碧浪中: 總論明人的臺灣認識 (Ming Conceptions of Taiwan: From Silhouettes to Islands)." *Tai da li shi xue bao* 臺大歷史學報 *(Historical Inquiry)* 40: 93–148.

Chou, Wan-yao 周婉窈. 2012. "Ming qing wen xian zhong tai wan fei ming ban tu li zheng 明清文獻中「臺灣非明版圖」例證 (An Illustration That "Taiwan Is Not Part of the Ming Dynasty Territory" in the Literature of the Ming and Qing Dynasties)." In *Hai yang yu zhi min di tai wan lun ji* 海洋與殖民地臺灣論集 *(A Collection of Essays on Maritime Taiwan and Colonial Taiwan)*, edited by Wan-yao Chou, 151–188. Taipei: Linking (聯經).

of "engineer" or "civil engineer" became *gong cheng shi* (工程師). It drew on the term *gong cheng*, which dated back more than a thousand years but appeared rarely until it was used during the Ming dynasty to describe the building procedures in flood control. *Gong cheng* ultimately came to be the closest equivalent to "engineering" in Chinese (Shih 2021).

Chu, Samuel C. 1963. "Liu Ming-ch'uan and Modernization of Taiwan." *The Journal of Asian Studies* 23(1): 37–53.

Davidson, James Wheeler. 1903. *The Island of Formosa, Past and Present: History, People, Resources, and Commercial Prospects. Tea, Camphor, Sugar, Gold, Coal, Sulphur, Economical Plants, and other Productions*. London and New York: Macmillan & Co.

DGBAS, Directorate General of Budget, Accounting and Statistics, Executive Yuan 行政院主計處. 2023. *Commonly Used Data for National Income Statistics*. Taipei: DGBAS.

Downey, Gary Lee. 2008. "The Engineering Cultures Syllabus as Formation Narrative: Critical Participation in Engineering Education through Problem Definition." *St. Thomas Law Journal (special symposium issue on professional identity in law, medicine, and engineering)* 5(2): 101–130.

Downey, Gary Lee. 2012. "The Local Engineer: Normative Holism in Engineering Formation." In *Engineering, Development and Philosophy: Chinese, American, and European Perspectives*, edited by Steen Hyldgaard Christensen, Carl Mitcham, Bocong Li and Yanming An, 233–251. Dordrecht: Springer.

Downey, Gary Lee. 2021. "Critical Participation: Taking on a Knowledge Hierarchy." In *Making and Doing: Engagement and Reflexive Learning in Science and Technology Studies*, edited by Gary Lee Downey and Teun Zuiderent-Jerak. Cambridge, 219–243. MA: MIT Press.

Faulkner, Wendy. 2009. "Doing Gender in Engineering Workplace Cultures II: Gender In/authenticity and the In/visibility Paradox." *Engineering Studies* 1(3): 169–189.

Finegan, Michael H. 1978. "Merchant Activities and Business Practices as Revealed in Several Manuscripts from Fukien." *Ch'ing-shih wen-t'i (Johns Hopkins University Press)* 3(9): 76–90.

Han, Kyonghee 한경희. 2021. *The Making of Korean Engineers: Historical Formation and Development* 한국 엔지니어의 형성과 발전. Paju: Dulnyeok (들녘).

HC iTaiwan Forum. 2019. "Zhao shi tao yi de tian zhi jiao zi wang huang san jiao shou '肇事逃逸'的天之驕子王晃三教授 (Professor Wang Hwan-san, the Proud Son Who Hit and Run)." https://www.youtube.com/watch?v=DNhPHMPK4GU. Accessed on May 27, 2024.

Hong, Sao-Yang 洪紹洋. 2016a. "Tai wan gong kuang gong si zhi min ying hua 台灣工礦公司之民營化: 以分廠出售為主的討論 (The Privatization of Taiwan Industrial and Mining Company: Focusing on the Selling of Individual Factories)." *Tai wan she hui yan jiu ji kan* 台灣社會研究季刊 *(Taiwan: A Radical Quarterly in Social Studies)* 104: 103–148.

Hong, Sao-Yang 洪紹洋. 2016b. "1950 nian dai tai ri jing ji guan xi de zhong qi yu tiao zheng 1950年代臺、日經濟關係的重啟與調整 (Re-establishment and Adjustment of Taiwan-Japan Economic Relations in 1950s)." *Tai wan shi yan jiu* 臺灣史研究 *(Taiwan Historical Research)* 23(2): 165–210.

Hsieh, Hsin-Chin. 2017. "History, Politics, and Identity: Joyce Huang and Taiwanese American Literature." *Chinese America: History and Perspectives*: 27–34.

Hsu, Hsueh-chi 許雪姬. 1993. *Man da ren zui hou de er shi nian: Yang wu yun dong yu jian sheng* 滿大人的最後二十年: 洋務運動與建省 *(The Last Twenty Years of the Mandarins: The Westernization Movement and the Founding of the Province)*. Taipei: Independence Evening Post (自立早報).

Hsueh, Jen-Yang 薛人仰. 1990. "Tai wan jiao yu zhi chong jian 臺灣教育之重建 (The Reconstruction of Education in Taiwan)." In *Kang zhan shi qi shou fu tai wan zhi zhong yao yan lun* 抗戰時期收復臺灣之重要言論 *(Selected Important Documents on Recovering Taiwan 1937–1945)*, edited by Jui-Cheng Chang 張瑞成, 299–304. Taipei: Party History Committee of the Central Committee of the Chinese Kuomintang (中國國民黨中央委員會黨史委員會).

Hu, Ding-Hua 胡定華. 2011. "Taiwanese IT Pioneers: Ding-Hua Hu." Interviewed and edited by Ling-Fei Lin 林玲妃. Mountain View, CA: Computer History Museum.

Jacoby, Neil H. 1967. *U.S. Aid to Taiwan: A Study of Foreign Aid, Self-help, and Development*. New York: F. A. Praeger.

Kao, Shu-Yuan 高淑媛. 2016. *Tai wan gong ye shi* 臺灣工業史 *(Taiwan Industrial History)*. Taipei: Wunan (五南).

Kennedy, Thomas L. 1974. "Self-strengthening: An Analysis Based on Some Recent Writings." *Ch'ing-shih wen-t'i* 3(1): 3–35.

Kirby, William C. 2010. "Engineers and the State in Modern China." In *Prospects for the Professions in China*, edited by William P. Alford, William Kirby and Kenneth Winston, 283–314. London: Routledge.

Kleine, Marie Stettler. 2020. "Proselytizing Problem-Solving: The Religious and Secular Values of Engineering for Good." Ph.D. Dissertation, Science, Technology, and Society, Virginia Tech.

Kuo, Wen-Hua 郭文華. 1996. "Tai wan yang wu ke ji chu tan: Cong tai wan tie lu, tai bei ji qi ju yu ji long mei kuang chu fa de chu bu tao lun 臺灣洋務科技初探(1887–1895): 從臺灣鐵路、台北機器局與基隆煤礦出發的初步討論 (Technology in Taiwan's Self-Strengthening Movement (1887–1895): A Preliminary Discussion of the Taipei Machinery Bureau, Taiwan Railway and Keelung Coal Mine)." *Xin shi xue* 新史學 *(New History)* 7(2): 99–138.

Kwong, Ki Chiu. 1887. *An English and Chinese Dictionary*. Shanghai: Wah Cheung.

Lamley, Harry J. 1964. "The Taiwan Literati and Early Japanese Rule, 1895–1915: A Study of Their Reactions to the Japanese Occupation and Subsequent Responses to Colonial Rule and Modernization." Ph.D. Dissertation, Far Eastern and Russian Institute, University of Washington.

Lamley, Harry J. 1981. "Subethnic Rivalry in the Ch'ing Period." In *The Anthropology of Taiwanese Society*, edited by Emily Martin Ahern and Hill Gates, 282–318. Stanford, CA: Stanford University Press.

Lin, Ching-Po 林清波, ed. 2012. *Tai wan ying zao ye bai nian shi* 臺灣營造業百年史 *(Taiwan's Construction Industry: A Centenary History)*. Taipei: Yuan-Liou (遠流).

Lin, Hsiao-ting 林孝庭. 2021. *Jiang jing guo de tai wan shi dai* 蔣經國的台灣時代: 中華民國與冷戰下的台灣 *(The Chiang Ching-Kuo Era: The Republic of China on Taiwan in the Cold War)*. Taipei: Walkers Cultural Enterprises (遠足).

Lin, Man-horng 林滿紅. 2008. *Cha, tang, zhang nao ye yu tai wan zhi she hui jing ji bian qian (1860–1895)* 茶，糖，樟腦業與臺灣之社會經濟變遷 *(1860–1895) (Tea, Sugar, Camphor Industry and Socioeconomic Changes in Taiwan (1860–1895))*. Taipei: Linking (聯經).

Lin, Mei-Ling 林美玲. 1996. "Bu zheng chang' de jiao da xue sheng 「不正常」的交大學生 ('Abnormal' Jiao Da Students: Abnormal Club)." *Jiao da you sheng* 交大友聲 *(The Alumni Voice of Chiao Tung University)* 356: 53.

Lin, Yi-Ping, and Hsin-Hsing Chen. 2021. "STS on the Street and at the Court: Interlocutors in the Taiwan RCA Collective Occupational Disease Lawsuit." In *Making and Doing: Activating STS through Knowledge Expression and Trave*, edited by Gary Lee Downey and Teun Zuiderent-Jerak, 37–50. Cambridge, MA: MIT Press.

Liu, Shinkei 劉進慶. 2012 [1992]. *Tai wan zhan hou jing ji fen xi* 台灣戰後經濟分析 *(Taiwan Postwar Economic Analysis)*. Taipei: Renjien (人間).

Liu, Ming-Chuan 劉銘傳. 1958. *Liu zhuang su gong zou yi* 劉狀肅公奏議 *(Liu Ming-Chuan's Proposals to the Qing Court)*. In *Taiwan wen xian cong kan*臺灣文獻叢刊 *(Taiwan Literature Series)* Vol. 27, edited by Bank of Taiwan Economic Research Office. Taipei: Bank of Taiwan.

Lo, Ming-Cheng M. 2002. *Doctors within Borders: Profession, Ethnicity, and Modernity in Colonial Taiwan*. Berkeley, CA: University of California Press.

Lucena, Juan, Jen Schneider, and Jon A. Leydens. 2010. *Engineering and Sustainable Community Development*. San Rafael, CA: Morgan & Claypool Publishers.

McCausland, Shane. 2011. *Zhao Mengfu: Calligraphy and Painting for Khubilai's China*. Hong Kong: Hong Kong University Press.

Norberg, John. 1999. *Three Tigers and Purdue: Stories of Korea, Hong Kong, Taiwan, and an American University*. West Lafayette, IN: Purdue University.

Ou, Su-Ying 歐素瑛. 2002. "Zhan hou chu qi taiwan zhong deng xue xiao zhi xue feng yu xun yu (1945–1990) 戰後初期臺灣中等學校之學風與訓育 (1945–1949) (Campus Atmosphere in Middle School vs. Disciplinary Education (1945–1949))." *Guo shi guan guan kan* 國史館館刊 *(Bulletin of Academia Historica)* 2: 1–41.

Ownby, David. 1990. "The Ethnic Feud in Qing Taiwan: What is this Violence Business, Anyway? An Interpretation of the 1782 Zhang-Quan *Xiedou*." *Late Imperial China* 11(1): 75–98.

Saxenian, AnnaLee. 2006. *The New Argonauts: Regional Advantage in a Global Economy*. Cambridge, MA: Harvard University Press.

Shih, Bono Po-Jen. 2021. "Towards an Engineering Ethics with Non-engineers: How Western Engineering Ethics May Learn from Taiwan." In *Engineering and Philosophy: Reimagining Technology and Social Progress*, edited by Zachary Pirtle, David Tomblin, Guru Madhavan, 161–180. Cham, Switzerland: Springer Nature.

Shih, Bono Po-Jen. 2022. "Engineering as Technology of Technology and the Subjugated Practice." Ph.D. Dissertation, Science, Technology, and Society, Virginia Tech.

Smith, Jessica M. 2021. *Extracting Accountability: Engineers and Corporate Social Responsibility*. Cambridge, MA: MIT Press.

Speidel, William M. 1967. *Liu Ming-ch'uan in Taiwan, 1884–1891*. Ph.D. Dissertation, History, Yale University.

Speidel, William M. 1976. "The Administrative and Fiscal Reforms of Liu Ming-ch'uan in Taiwan, 1884–1891: Foundation for Self-strengthening." *The Journal of Asian Studies* 35(3): 441–459.

Wang, Chih-Jen 王志仁. 1997. "Cao xing cheng li pin ban dao ti ba zhu 曹興誠力拚半導體霸主 (Tsao Shin-Chen Goes All Out to be the Boss of the Semiconductor Industry)." Tian xia za zhi 天下雜誌 *(CommonWealth Magazine)* 197: 53–62.

Wang, Li-Chun 王麗娟. 2013. *Shi min yu shu wei shi dai de gu shi* 施敏與數位時代的故事 *(Simon M. Sze and the Stories of the Digital Era)*. Taipei: Wa-People (紅津).

Wang, Shan-Shan 王珊珊. 2004. *Jin dai Taiwan zong guan tie lu yu huo wu yun shu zhi yan jiu (1887–1935)* 近代台灣縱貫鐵路與貨物運輸之研究 *(1887–1935) (A Study on Taiwan's Railway and Cargo Transportation in Modern Times (1887–1935))*. Hsinchu, Taiwan: Bureau of Cultural Affair, Hsinchu County Government.

Wisnioski, Matthew, Eric S. Hintz, and Marie Stettler Kleine, eds. 2019. *Does America Need More Innovators?* Cambridge, MA: The MIT Press.

Wu, Shu-Min 吳淑敏. 2019. *Hu ding hua chuang xin xing zhuan* 胡定華創新行傳 *(Hu Ding-Hua and the Acts of His Innovation)*. Hsinchu, Taiwan: Rehoboth Pioneer Networking (力和博).

Wu, Tsong-Min 吳聰敏. 2023. *Tai wan jing ji si bai nian* 台灣經濟四百年 *(A 400-Year History of Taiwanese Economy)*. Taipei: Spring Hill (春山).

Yang, Meng-Hsuan Dominic, and Mau-Kuei Chang. 2010. "Understanding the Nuances of 'Waishengren': History and Agency." *China Perspectives* 83: 108–122.

Yap, Ko-Hua 葉高華. 2021. "Cong jie mi dang an zhong gu er zhan hou yi ru tai wan de wai sheng ji ren shu 從解密檔案重估二戰後移入臺灣的外省籍人數 (Reassessing Number of Mainland Chinese Immigrants with Declassified Archival Data)." *Tai wan shi yan jiu* 臺灣史研究 *(Taiwan Historical Research)* 28(3): 211–229.

York, Emily. 2018. "Doing STS in STEM Spaces: Experiments in Critical Participation." *Engineering Studies* 10(1): 66–84.

Yue, Meng. 1999. "Hybrid Science versus Modernity: The Practice of the Jiangnan Arsenal, 1864–1897." *East Asian Science, Technology, and Medicine* 16: 13–52.

2 Freelancing Around Japanese Infrastructures

Chu Jiang-Huai (朱江淮, 1904–1995) climbed. In the 1920s, roughly thirty years into the colonial period, he had travelled to Japan with family support at age twenty-two to earn a bachelor's degree in electrical engineering at Kyoto Imperial University. The first local electrical engineer in the island's history, Chu secured a prestigious job at Taiwan Electric Company, the government-run electric utility that was then building a massive hydroelectric power station. The knowledge Chu gained in the engineering sciences on the Japanese mainland marked him as an elite technical talent (技術者) in the colony, associating him with Japan and its powerful competitors in the West. But Chu was not hired into an engineering department. Local engineers were expected not to compete with comparably-trained Japanese engineers. In 1930, Chu started out in the sales department (業務部). Even though his title was an assistant engineer (技手) in the department's engineering section (技術課), he was assigned to help persuade local people in their native language to expand their use of electricity (Chu 2003, 69–75).[1]

Education at Kyoto did give Chu more options than the vast majority of local men who sought to work in industry on the island. And he took advantage. Remarkably, he was able to climb up within Taiwan Electric to become its professional engineer in 1941, during what Japan called the Greater East Asia War (Chu 2003, 90). Chu also used his unique status to help other locals gain access to the infrastructures of education and work that the Japanese Governor-General's office, the *Sōtokufu*, made available to them.

In 1930, the year he graduated, Chu took a dramatic step. He joined with eleven other men with education in engineering and the sciences to found the Taiwan Polytechnic

[1] The term "native language" oversimplifies the dialects it represents. During the late Qing period, different provinces and large cities spoke different dialects of Mandarin. The dominant local dialect spoken in Taiwan was close to Hoklo. When the KMT gained control of Taiwan in 1945, they called it Hoklo. It became known as Taiwanese after the 1980s. For our purposes here, it is important only that Chu could communicate with locals and Japanese engineers and managers could not.

K.-H. Chang et al., *Engineers and the Two Taiwans*, Synthesis Lectures on Global Engineering, https://doi.org/10.1007/978-3-031-59766-4_2

Fig. 2.1 Taiwan Polytechnic Society (front right first—Chu Jiang-Huai; second—then-president Tu Tsung-ming; third – Wang Chao-Ing) (Courtesy of Mr. Chu Ray-Yong)

Society. The Society's name was aspirational, a claim that the island could be or become a thing unto itself, even as a colony within the empire. The Japanese called the local people "natives," and prior to the colonial period locals had been residents of a province of Qing China. Chu later remembered that "The Japanese practiced a discriminatory policy on us. Most of us young Taiwanese were upset and felt we needed to unite ourselves to contribute to our land Taiwan." In 1930, the strategy was to "encourag[e]" more students to study "science, engineering, agriculture, and mining" and make "industrialization" on the island about something more than supporting the Japanese empire (Chu 1980a, b, T. Lin 1951) (Fig. 2.1).

Wang Li-Chung (王禮忠, 1918–?) also sought to climb, but within private Japanese companies rather than those run by the government. In the 1930s, his father somehow found the money to send him to the industrial school in Taipei to pursue a degree in applied chemistry. Sixteen years old, he had spent two extra years in elementary school for local children and passed the industrial school's entrance exam on the third try. After completing his degree, Wang found work supporting Japanese technicians in a sugar refinery in Tainan to the south, and then in an alkali factory back home in Taipei.

In contrast with Chu, Wang was not able to climb far nor was he positioned to imagine Taiwan as a thing unto itself. In 2006, he reported in an interview, "[W]e were resigned to the fact" that the workplace was "discriminatory" because for "400 years... [w]e didn't have our own legitimate country." He was glad to get as far as he did, building personal connections with Japanese students while at the industrial school. Indeed, after retirement Wang organized reunions of former local and Japanese classmates who wanted to see one another again, arranging gatherings in what had become distinct places, Taiwan and Japan (Cheng and Yang 2009, 31–44).

Li Shui-Pien (李水變, 1910–?) sought a different kind of pathway into Japanese industrial infrastructures from either Chu or Wang—outside of Japanese companies. After completing studies at the relatively new industrial school in Tainan at age 28, he founded a small machine tool company. At the time, jobs for local graduates of the school were scarce on the island. Some locals sought to climb within Japanese companies by finding positions in Manchukuo, the territory Japan had seized in northeast China in 1932 (H. Hsu 2002, Kao 2011, 32–43). There their schooling and Japanese language skills put them at an advantage over Chinese locals. Li Shui-Pien stayed in Taiwan, however, using contacts with former Japanese classmates who worked in nearby sugar refineries to secure orders for making and repairing components for their machinery. By running his own company, Li was able to avoid working within a Japanese company and remain at home on the island (Kao 2016, 168–170, 2011, 33–35) (Fig. 2.2).

The episode in this chapter follows the trajectories of would-be engineers and other industrial workers in Taiwan who sought to climb the engineering infrastructures of Japanese colonial governance. The Japanese understood engineers as technical talents who had knowledge of machinery and worked to build "industry," meaning modes of economic production that included machinery (Kao 2016, 116–118, Han 2021). During the 50-year colonial period, the engineers who worked on the island nearly always came from the Japanese mainland where they earned degrees in engineering through curricula that drew from precedents in Great Britain and Germany. The Governor-General's office in the colony worked with Japanese companies to attract as many engineers as would come, as well as industrial technicians to support them (Huang 2012, Kao 2016, 116, Yanaihara 1999, 51).

The colonial government did build infrastructures of educational formation and technical work for local boys and men to provide pathways for them into Japanese industry. Such pathways were attractive because most locals who went to work in new industrial arenas had to join as manual laborers (P. Hsu 2013, C.Y. Tu 1993, 138–142). Employers in both private and government-owned industries filled in gaps around Japanese technicians with local men who had demonstrated the means and initiative to gain the machine knowledge needed to climb above manual labor, usually through education (Cheng 2012, 34–43, 76–77). But these locals then encountered glass ceilings, as Wang Li-Chung did when he became an assistant technician in a private sugar refinery.

In the colony, congruences between the geographical identities of technical workers and their technical identities quickly produced a dominant pattern. Those workers who grew up on the island had to plot career pathways subordinate to those who came with equal training from the Japanese mainland. Certainly, the substantial differences that had

Fig. 2.2 Taipei Industrial School, 1934 (Courtesy of National Taipei University of Technology)

emerged among locals during the late-Qing period did not disappear, e.g., differences among those with dramatically different levels of income or those who traced family histories from different provinces. But now, all inhabitants of the island, including indigenous peoples, had to accept a new geographical identity, that of "subject" of the Japanese emperor. And with that identity came a hierarchy between "colonist" subjects from the mainland and "colonized" subjects in Taiwan that infused the organizations of industry.

Yet, as we will see, the meaning of that hierarchy for local technicians and, later, engineers evolved over time as many also came to champion their connection to Taiwan. In particular, by the 1930s an increasing number of would-be engineers sought pathways into industry akin to what Li Shui-Pien followed, establishing their own companies. Those who did became what locals came to call "freelance" workers, who benefitted from a freedom to move akin to that of doctors (C. Lin 2007, Cheng 2012, Liu and Lin 2008). The new industrial freelancers hired local labor and elevated themselves via contractual relationships with the colonial government and private Japanese companies (C. Lin 2007). In so doing, they extended practices initiated by gentry merchants during the Qing era who had used contracts with the government to elevate themselves. The difference this time was that freelancing companies were not only trading agricultural goods but also finding ways into the infrastructures of machine-based industry. And by establishing small gaps between themselves and Japanese companies, they were able to build connections with local workers and the island that were beyond direct Japanese supervision and control.

Common School Knowledge for the Underside of Industry

Despite having no experience managing a colony and relatively little funding to spare, the Governor-General's office quickly went to work extending railroad lines, improving the Keelung harbor that served Taipei, building administrative offices, simplifying land-holding arrangements to facilitate population counts and tax collection, and establishing government monopolies on the production and sale of salt, camphor, tobacco, and opium (H. Lin 1977, 615). The colonial government also offered subsidies to private companies to begin adding machinery to agricultural production, especially by building refineries to process cane sugar.

To provide managerial labor for its projects, the government attracted newly-trained talents from the Japanese mainland through handsome supplements of as much as 60% of salaries or wages or promotions to higher ranked positions (Cheng 2009, 102, Huang 2012, 33). Indeed, during the first decade nearly 60,000 skilled workers, managers, and administrators came, including industrial technicians and engineers (Tsurumi 1977, 53). Although the former Qing provincial authority had trained several dozen technicians at the Hall of Western Learning, we could find no evidence that Japanese employers made use of them (Kuo 1996, 123–124).

Japanese administrators did develop a system of basic elementary schools, called "common" schools, in part to provide unskilled workers for colonial projects. With language training in Japanese a top priority, common schools drew upon educational practices that the Meiji government had built over the previous two decades to foster loyalty to the emperor and commitment to Japan's industrial development. Local boys were to recognize and understand that the Meiji emperor had taken the mandate of heaven away from the Qing emperor, and to appreciate that the availability of schooling at all demonstrated the emperor's great concern for his subjects in Taiwan. Commitment to industrial development was taught as a moral responsibility. Through much of the colonial period, for example, local boys at the schools completed an ethics lesson about a great Japanese weaver (Tsurumi 1977, 144–145). The purpose of the lesson was not to highlight the weaver's technical skill but to explain how his town had become a great industrial center. Pupils at common schools were to respect Japanese machinery and industrial power even through a technical practice that had long been associated with women (Fig. 2.3).

As another lesson elaborated, young boys were to fulfill their duty by earning a living and gaining new knowledge necessary to serve the empire. Their first option was to follow the occupations of their fathers and grandfathers, including artisanal work and running

Fig. 2.3 Common school, 1929 (Courtesy of National Museum of Taiwan History)

small businesses, but now using new knowledge and techniques to do so more efficiently. Still other lessons pointed pupils specifically toward manual work in the new industries (P. Hsu 2014). While colonist boys and, later, girls attending primary schools built for them were to aspire to achieve mobility for themselves and their empire, colonized boys and, later, girls were not. They were to accept serving from the bottom.

A 1912 memorandum from the Governor-General's office to other officials outlined how Japanese administrators understood the task of educating natives. As Tsurumi (1977, 51) put it, the colonial government saw locals as "frivolous, fond of idle shows and foppishness, interested only in the hollow phrases of Chinese learning." They "despised manual labor and had nothing but contempt for honest and useful occupations which required physical skills or exertion." In other words, the colonial government was likely confronting the hierarchy of value that had bedeviled Liu Ming-Chuan during his term as Qing governor in Taiwan. For Japanese administrators, the common schools had to be attractive enough to persuade local families to enroll their children and prepare them for work, but not so attractive as to raise expectations of upward mobility and leadership. They had to fit the intellectual level of subordinate natives whose "education needs," as officials rationalized, were "not the same as those of the Japanese" (cited in Tsurumi 1977, 69).

Along with the Japanese language, the main strategy for student learning was to emphasize arithmetic and military gymnastics. When the Japanese military had gained control in 1895, many residents accepted its offer to leave the island, usually by relocating to Fujian province across the strait. The act of staying behind therefore conveyed the presumption of acceptance of imperial authority, even if the reason for staying was the cost of moving or the absence of family connections. In like fashion, learning Japanese, arithmetic, and gymnastics came to presume an interest in jobs with arriving Japanese employers. Arithmetic in particular opened a door to worlds of calculation, including science, that Euroamericans valued and that Japanese companies brought with them. And in 1904, common school knowledge added the "basics of craftsmanship, agriculture, marine products, and business [commerce] as an option" (Li 2005, 75). Attendance at common schools actually rose quickly from roughly 10,000 pupils in 1899 to more than 900,000 at the end of the colonial period (P. Hsu 2014, 135). To meet what became rapidly-growing demand, the Governor-General's office often required wealthy local families, usually those of merchants and large landholders, to contribute funding before it agreed to establish a school (Komagome 2019, 255–263).

Because the island had become part of the emperor's realm, wealthy families could actually send children to the Japanese mainland for schooling. If local children could demonstrate proficiency in Japanese and meet entrance requirements, they could attend seven-year higher schools (旧制高等学校), and, for a precious few, three or four years at a university (S. Hsu 2009, 17). Because the language barrier proved to be too great at first, accomplished common school graduates from wealthy families tended to make their way into schools for teacher training or training in Western medicine. Private Chinese schools did remain open throughout most of the colonial period, but the colonial government undermined the status attributed to classical Chinese learning and attendance declined (P. Hsu 2014, 136). The most realistic option for learning by local children was common school knowledge that pointed them to the underside of industry.

Industrial Knowledge to Climb Up ... or Out?

For the first half of the colonial era, a period of twenty-five years, industrial schooling—learning to handle machinery—was largely off-limits to local youth, with one exception that would become increasingly significant over time. In 1912, Japanese administrators established in Taipei an educational infrastructure with the formal mission to "develop higher industrial students as the primary goal" for local boys (Cheng 2012, 16). It was a bit of an overstatement.

Many of the technicians who arrived from Japan had gained their knowledge of machinery in a five-year curriculum at one of 36 industrial schools. They learned a significant amount of mechanics and electrical theory in machinery operation. In contrast, the new Institute of Industrial Instruction for the colony offered local adolescent boys only three years of learning in two tracks, learning that was mostly training (Cheng 2012, 25–26). The woodworking track included general woodworking and furniture-making, while the metalworking and electrical track channeled students into casting, forging, metal processing, sheet metal, and electrical work (Li 2005, 238). A mechanical track would be added five years later. In the school's first year, eleven Japanese instructors led 58 students (Cheng 2012, 23).

Despite the low level of technical education, it is likely that students came from wealthy families. A later observer calculated that the annual cost of attendance was high, nearly equal to the annual salary of a local teacher at a common school (Cheng 2012, 105). Still, during the Institute's first decade, attendance at the school would grow to exceed 300, led by more than 40 instructors (Cheng 2012, 65). The promise of the school was the opportunity to climb high enough to assist Japanese technicians.

Many graduates went to work in the rapidly-expanding sugar industry in which mechanized sugar processing replaced the open-kettle boiling of cane juice with closed vessels heated by steam. Adding machinery made it possible to put the vessels under a partial vacuum, enabling boiling at lower temperatures, saving fuel and increasing throughput. Between 1902 and 1918, mechanized refining increased from 2% to 87% of total sugar production, with tonnage increasing from one to more than 300 tons (Liao 1993) (Wang 2011). World War I then dramatically increased demand from Europe where cane sugar was replacing European beet sugar (Fig. 2.4).

Fig. 2.4 Sugar cane mill in the Japanese colonial period (Courtesy of National Museum of Taiwan History)

By 1918, food processing factories accounted for more than half of all industrial production in the colony, with sugar refineries amounting to roughly half of that (Liao 1993, 17). In that year, more than 26,000 workers labored in the food processing factories, followed by roughly 6,000 in chemical companies, 2,000 and 1,000 in machine tool and textile companies, and 4,000 distributed across other industries (Liao 1993, 20). At a time (1920) when roughly 1.2 million people still remained in local communities to work in agriculture, mostly as peasants or wage laborers, graduates of the Institute for Industrial Instruction traveled across the island to find more attractive work in these industries.

But as we saw above with Li Shui-Pien, not all Institute graduates were satisfied with climbing within Japanese companies. Lin Shih-Tsao (林煶灶1893–1971) was one of the first graduates from the woodworking track. After working for four years in Japanese construction firms, he joined together with two local friends to form their own company, the Xie-Zhi Business (協志商會). Xie-Zhi subcontracted work from the large firms, supplying raw materials, supporting workers' needs, and offering a range of logistical support. Akin to the late-Qing gentry merchants, the company positioned itself as a mediator between the local production knowledge of local providers and, now, the industrialized practices of large government and government-supported enterprises. Over time, the company was able to acquire sufficient machine knowledge to bid on construction jobs itself, formally joining the Japanese construction industry in competition with other members. To build a reputation and solidify its position, the company deliberately underbid Japanese firms and then spent as much as twice the bid price to complete the work (Cheng 2012, 82–87).

A 1940 survey of graduates from the Institute of Industrial Instruction found that one-third of them owned their own businesses (Cheng 2012). Other graduates founded what became the Guang-Zhi Business (光智商會) and the Tang Eng Iron Works (唐榮鐵工廠). It is not possible to know just how many Institute graduates tried and failed to establish going concerns because they could not identify the right gaps or could not attract enough financial support to weather the challenges of start-up. It is clear, though, that establishing one's own company in an expanding industrialized and militarized environment controlled by colonial rulers provided a kind of escape (Cheng 2012, 214–216, Kao 2016, 157–179). At a time when political liberation was virtually impossible to even imagine, becoming a freelancing entrepreneur meant that one could find spaces beyond direct Japanese control and prioritize attachments to other locals and the island.

Spectators of Industrial Expansion

In the late 1920s, a civilian Governor-General exclaimed excitedly that industrial development was transporting the colony into a new era. Continuing to rely on agriculture without machinery, he asserted, would eventually have reached a "bottleneck," limiting further expansion (Wang 2011, 58). But agriculture with machinery was not only making the island self-sufficient, it was also turning the island into a source of manufactured exports into south China and Nanyang (Southeast Asia). Economic production through machinery, he imagined, was unbounded. During the next decade, that assessment appeared to be coming true as industrial successes in the colony supported emerging political plans on the mainland to expand the empire dramatically to the south and east. The expansionist vision of an empire led by a benevolent ruler would gain a distinctive name—the Greater East Asia Co-Prosperity Sphere (Myers and Peattie 1984, 123).

As work on the Keelung Harbor was being completed during the 1920s, two massive projects to support industrial development were gearing up. The Chianan reservoir and irrigation project would expand the arable area in the south from about 12,000 acres to more than 370,000 acres. Completed in 1930, the massive and predictable water supply it provided also increased the number of harvests from two to three a year. With completion of the irrigation project, the Governor-General's office began to pressure local farmers, who owned roughly 80% of the farmland, to shift production away from food crops for local consumption, especially rice. A supporting move on the mainland was to impose controls on the import of cheap Taiwanese rice (C.Y. Tu 1993, 99–100, Chang 1980, 32). In order to fulfill their responsibility to the emperor, farmers were now to shift to cash crops, especially sugar cane, that would feed raw materials to Japanese-owned factories and facilitate exports. Upper-class landholders and owners of small factories were to buy and implement Japanese machinery as much as possible, which would enable them to gain the financial benefits of increased outputs and the symbolic benefits of official approval (Kao 2016, 116–117) (Fig. 2.5).

Machines needed electricity to function. The huge Sun-Moon Lake hydroelectric plant came on line in central Taiwan in 1934, delivering 150 MW across the island with plans to eventually distribute 2,000 MW. It had taken 15 years to build, including a five-year hiatus after World War I. The massive infusion of cheap electricity warranted expanded attention to other industries beyond agriculture, especially the chemical industry, and the import of all kinds of Japanese machinery to appropriate and exploit them.

The colony could now also import raw materials from Nanyang and South China to factories producing a growing range of industrial and consumer products (Kao 2016, 115–125, Liao 1993, 38). Kaohsiung in the south, for example, became a center for the production of aluminum, a lightweight metal with a wide range of applications, including military hardware. Although the island had no bauxite to mine, it now had the capability to smelt, extracting the ore by heating it to high temperatures, melting it, and reducing it chemically. By importing bauxite from Malaya to the new port city, Japanese aluminum companies in the colony came to supply roughly half of the mainland's demands for the metal (Chang 1980, 80). New chemical companies proliferated across the island, producing such goods as fertilizers, sodium carbonate (for glass), ethanol, castor oil (for feedstocks), paper, and synthetic fibers, as well as such consumer goods as hemp, spices, medicine, cassava, starch, and paint (Kao 2016, 158–166).

Machines also needed knowledgeable technicians to run them. The mainland could supply technicians only up to a certain level of flow, and domestic opportunities there

Fig. 2.5 Chianan reservoir and irrigation project, 1930 (Courtesy of National Museum of Taiwan History)

were also expanding. The colonial government still judged native boys to be largely incapable of acquiring and performing the advanced machine and organizational knowledge necessary to manage manufacturing machinery and manual workers. It thus turned its attention to the growing communities of colonists that were producing a new generation of Japanese boys for whom industrial knowledge and associated occupations would indeed be appropriate.

In 1918, the Institute for Industrial Education in Taipei introduced a five-year curriculum for mainland boys to prepare them to serve as industrial technicians, leaving in place the three-year curriculum for locals. A year later the Office of the Governor-General upgraded the institute to the Taiwan Public Taipei Industrial School, equal to industrial schools on the mainland (Cheng 2012, 2, Li 2005, 326). The new school-level curriculum matched tracks of learning to the expanding range of industrial work practices. Different categories of mechanical machinery, for example, expected to draw on distinct sources of mechanical power, e.g., coal-fired furnaces, steam engines, gasoline, and hydroelectric power, and all depended upon operators who possessed some knowledge of electricity (Cheng 2009, 87). At the outset, the curricular tracks included mechanical, chemical, and electrical industrial production as well as civil and architectural construction. And the specific configurations of knowledge and learning that made up each curricular track were designed to produce distinct types of technicians.

The mechanical machinery track, for example, began with two years of courses shared with other tracks. These included introductions to chemistry, physics, mechanics, and cartography, as well as mathematics courses in algebra, geometry, and trigonometry (Tainan Junior College of Engineering 1934, 13–19). By 1935, courses that included calculation were complemented with "exercise" courses in which students practiced solving problems under the supervision of assistants. More advanced courses then introduced students to specific categories of machines, such as steam machines, motors, engines, refrigeration systems, water turbines, and other machinery used on ships and in textile mills. Field practice and experimentation with actual machinery took place exclusively on campus. Notably, technician knowledge did not include calculus, differential equations, or advanced mechanical theory (e.g., dynamics)—all of these were appropriate only for engineers trained on the mainland (Taipei Industrial School 1935, 10).

The stark divisions between colonist and colonized subjects in industrial education and work persisted even amidst a formal administrative move to eliminate them. In 1922, the Japanese government gave in to persistent calls, requests, and demands for more equitable educational opportunities, especially beyond the elementary level. It declared all schools, including Japanese primary schools, as equally accessible to both colonists and natives. Local boys and girls could compete with Japanese boys and girls for admission to any school. But there was a hitch: all seats would go to the most qualified applicants. With Japanese language skills the most important criterion, few local applicants were able to gain admission to formerly Japanese-only schools. In addition, the new policy actually increased pressure on local students and their families to cultivate and demonstrate *kokumin seishen*, i.e., "national spirit."

The Taipei Industrial Institute implemented the policy by eliminating its three-year track, thus granting Japanese applicants a clear edge in admissions. Incoming classes at the school during the second half of the colonial period ranged from 300 to 1000 students. The classes included between 25% and 75% of the Japanese boys who completed the entrance examination and only 15–30% of those local boys who attempted the same (Cheng 2009, 92–93).

In the late 1920s, the Governor-General's office approved a second industrial school, this time in the south. Its twin, overlapping goals were to help develop growing industries across the central and southern regions of the island and to support what had become known among the Japanese as the Southern Expansion of the empire (Wang 2011, Kao 2002). Since the focus of the Tainan Higher Industrial School (臺南高等工業學校) would not include transportation and communication systems, its curriculum would contrast with the Taipei school by not including civil or architectural engineering. Opened in 1931, the new school featured a chemical industry department (應用化學科) along with departments dedicated to the study of mechanical machinery (機械工學) and electrical machinery (電氣工學) (Wang 2012, 26). The sugar industry had remained dominant in the south, seeking a regular supply of chemical technicians. Administrators of the colony

Fig. 2.6 Tainan Higher Industrial School, 1939 and its first president Wakatsuki Michitaka (Courtesy of National Taiwan University Library)

and empire worried about possible interruptions in oil supplies and called for the production of ethanol as a substitute. Mechanical and electrical machinery students were to learn to build, fix, and maintain industrial machinery, especially for sugar refining, refrigeration, and air conditioning. Continued expansion of chemical industries across the region justified adding a department of electrical chemistry in 1940 (Fig. 2.6).

The Tainan Higher Industrial School initially recruited three engineering graduates from the University of Tokyo to serve as professors. Three associate professors were graduates of industrial schools on the mainland, and two adjunct lecturers came to the school from industry. All were Japanese. To support them, each department also hired assistant technicians and laboratory assistants, some of whom were locals.

By the 1930s, the knowledge of technicians had gained substantially in content and status relative to the knowledge taught in common schools. The curricula at Tainan added required courses in English and German to make use of imported Western textbooks. Hands-on encounters with machines moved off-campus to become two-month internships in the workplace. Also, to facilitate flows of students from educational to work infrastructures, the Tainan curricula supplemented machine knowledge with introductory levels of factory knowledge. For budding technicians, factory knowledge included industrial materials, industrial economics, factory management, factory construction, and factory hygiene (Tainan Junior College of Engineering 1934, 15–17).

As had been the case in Taipei, the Tainan school showed a clear preference for Japanese students. New dormitories were available only to them, suggesting that school administrators used the dorms to attract applicants from the north. Native students had to live off-campus. Presumably they were either local or could make local connections in a local-dominated city. By the end of the Greater East Asia War, only 109 of the school's 752 graduates were locals (Wang 2012, 44). One of the last local students admitted to the Department of Mechanical Engineering during the colonial period was Lin Chang-Cheng (林長城, 1923–2006), later founder of TECO Electric and Machinery. He reported that his department had 80 students, of whom only seven were locals (Chen and Kuo 1991, 189).

Even those locals who earned diplomas at one of the industrial schools encountered systematic inequities on the job (Kao 2002, 251–252). In addition to racial/ethnic discrimination against natives, Japanese companies had an additional reason to think twice before hiring local graduates. On the mainland, the *zaibatsu* – large multi-industry conglomerates—had begun implementing a practice of lifetime employment, thus cementing their status as family-like industrial households (from the samurai household, the "*ie*") (Downey and Wada 2011). In this infrastructural formation, new employees joined a company household akin to being born into a family household, measuring their age from the date of entry and moving from role to role as needed to serve the household and ensure its continuity. In this context, it was likely problematic for a Japanese company to grant a local graduate permanent membership in the corporate household. Those local graduates who aspired to and succeeded in gaining entry to Japanese organizations found that, when they did indeed become household members, they had to accept subordinated status as junior members.

For example, one Taipei graduate who wanted to find a job at Taiwan Electric Company later described how he would receive a monthly salary that was roughly 60% of that received by Japanese employees who had the same educational credentials and joined the Company's household at the same time. "I didn't think my grades were lower than Japanese," he said, "and receiving lower pay made me disgusted." Another local Taipei graduate at Taiwan Electric found out on the job that Japanese workers on survey crews received travel bonuses that were not available to him. "I could not bear this," he later reported (Cheng 2012, 79, Cheng and Yang 2009, 130, 144).

Both actually declined to pursue positions that local and Japanese peers might have otherwise seen as prestigious jobs with a government-owned company. They chose to go "overseas" to Manchukuo where no salary differential applied to them and where their language skills and technician training positioned them as quasi-colonial industrial experts across an environment populated with speakers of Mandarin Chinese, Manchu, and Mongolian who mostly lacked industrial knowledge (Kao 2011, 42).

Most Japanese graduates from Tainan easily found jobs in large Japanese private or government-owned companies. Some sought more advanced study, e.g., engineering degrees on the mainland. Japanese graduates were certainly free to create companies in the colony that employed local labor, but such could well have been seen in Japanese circles as a step down and few did. Some did connect with local colleagues to establish jointly-owned companies, mediating between Japanese customers and local workers. Meanwhile, native graduates typically found that they had to choose between going overseas and accepting lower-paid positions in smaller companies that were either partially or wholly owned by natives. In 1934, only 12 of the 32 natives who earned diplomas from the school received offers from Japanese companies (Kao 2011, 23) (Fig. 2.7).

Lai Tsai-Te (賴再得, 1919–1996), a native who became one of the first chemical industry graduates, was hired to solve a technical problem for a local-owned starch factory that served Japanese companies. Lai successfully figured out how to reduce the water content in the liquid starch while increasing its overall quality. The owner, realizing that his factory could benefit from the technician knowledge available at Tainan Higher Industrial School, persuaded three of his four sons to prepare themselves for admission to the school and subsequent careers as industrial technicians (Preparatory Working Group for Professor Lai Tsai-te Memorial Academic Symposium 1997).

Despite the success stories of local graduates, the total number of local boys who made their way through the Taipei and Tainan schools remained frustratingly small throughout the 1930s and into the 1940s. In both the north and the south, local men with industrial

Fig. 2.7 Local workers and a sugar mill press machine in the Japanese colonial period (Courtesy of National Museum of Taiwan History)

knowledge largely found themselves as spectators of Japanese industrial expansion. They were wholly welcome to participate only at the bottom in positions that did not merit lifetime employment.

Climbing Around Japanese Governance

As mentioned above, Chu Jiang-Huai co-founded the Taiwan Polytechnic Society with eleven other technical practitioners who had committed themselves to industrial development on the island. He was among the most elite of this privileged group with his degree from Japan in electrical engineering and new job at the government-owned utility company. The rationale, as he remembered decades later, was to attract more local students to study in the industrial fields of "science, engineering, agriculture, and mining." He said that "the smartest Taiwanese preferred to study medicine, law, or commerce" because these trajectories "could promise them freelance jobs in the future," i.e., beyond Japanese control (Chu 1980a, 5). And as we saw, gaining admission to the Taipei or Tainan industrial schools put most local students on career tracks that led to low-level employment within Japanese companies.

Leaders of the Taiwan Polytechnic Society positioned it as a "professional" organization. Embracing the idea of the professional was a decisive move in two ways. First, it enabled local graduates of the Japanese industrial schools to see themselves as connected to one another apart from their shared status as colonized subjects. As members of a profession, they had attained a shared technical identity beyond those assigned by Japanese employers. As Chu put it, the Polytechnic Society helped graduates "meet each other," "exchange their knowledge," and "build professional careers" (Chu 1980a, 5).

And second, membership in a professional society carrying the name "Taiwan" gave each of them a geographical identity serving the island and its people in ways other than as natives fulfilling their duty to the emperor. To claim to be a professional on the island was to embrace it as Taiwan, something more than a colony. And to do so as a polytechnic professional, which then meant an industrial professional, was to accept and assert industrialization as an integral condition of Taiwan's future.

Joining the Society thus affirmed that one could serve Taiwan's industrialization even through training at an industrial school that was conceived and run by Japanese and through employment within and around Japan's massive industrial infrastructure. The learning and work could be construed differently. Formally, the Taiwan Polytechnic Society was apolitical in the partisan sense that it did not position itself to challenge Japanese imperial power. Its members were well aware that they could not achieve legitimacy through conflict, let alone gain imperial permission to even exist. At the same time, the appearance and growth of the Taiwan Polytechnic Society during the 1930s did authorize members to assert, if only to themselves, that they worked for something other than expanding Japanese power.

Under the leadership of another core member of the Taiwan Polytechnic Society, Wang Chao-Ing (王超英; 1891–1971), the Polytechnic Society expanded beyond Tainan and Taipei. Wang worked in the only sugar factory run by a local owner. He had joined after completing his education at the Tokyo Higher Industrial School and then played a role in hiring additional local graduates to follow. Wang organized the founding of new chapters in Kaohsiung, an industrial center, and in Taichung, where a new industrial school focused on agriculture and forestry had opened (T. Tu 1970). Establishing those chapters had the effect of enabling the geographical identity of Taiwan to travel to two new cities. While not revolutionary in the conventional sense, these steps did perform a politics of resistance and self-determination.

During the last decade of the colonial period, an increasing ability of locals to claim a geographical identity in the name of Taiwan actually fit a growing tension in the empire's handling of the island colony. On the one side, the Japanese leadership attempted to pull the colony ever more tightly into a militarized vision of imperial expansion. Put most simply, the island was to become the empire's base for its southern expansion and realization as the Greater East Asia Co-Prosperity Sphere. In 1936, the mainland assigned a military governor to the colony, the first since the Governor-General's office had become a largely civilian operation in 1919. In 1937, the Japanese military used a local incident on the Chinese mainland to justify invading the continent. Beginning that year, new policies from the Governor-General's office emerged as marching orders.

Becoming a base for southern expansion meant an increase in centralized control within the empire. The island must not become a site of disharmony or discord. To that end, people on the island were given the opportunity to become Japanese; it was an offer from the emperor that they could not refuse. People must fully Japanize, accepting that all worked directly on behalf of the emperor, which now meant transforming the island into a base (Ching 2001). Most importantly, industrial production had to be redirected away from food and consumer goods and toward military ends. And in 1938, a revised policy of National Mobilization authorized the Governor-General to reduce the length of industrial schooling, narrow the curricular focus to training for more specific tasks, and limit the job choices of graduates to specific arrays of selections. The government also began granting hiring priorities to organizations and firms that needed technicians and assistant-technicians for war-related industries. It was important for the colony to become self-sufficient economically so it could continue to function as part of the empire even if it got cut off from the mainland.

At the same time, the pursuit of self-sufficiency carried an additional implication. By virtue of its prominence within the empire, Taiwan the island was gaining increasing separation from the Japanese mainland. With abundant and cheap electricity as well as developed ports to manage flows of raw materials and manufactured products, Japanese companies energetically expanded into such heavier industries as metal manufacturing, machine tools, ship repair, paper pulp, cement, and coal production. And following a five-year plan established in 1938, the Governor-General's office began founding industrial schools across the island, including in Hsinchu, Taichung, Changhua, and Chiayi, flowing north to south, and *Karenko* (Hualien) across the mountains on the east coast. In 1943, in the midst of the Greater East Asia War, it added a college of engineering to the imperial university it had established in Taipei. The mainland government planned to begin producing engineers within the colony rather than attracting them to the colony (The Department of Electrical Engineering at National Taiwan 2016) (Fig. 2.8).

The imperial government's war effort thus dramatically expanded opportunities for native industrial workers who were in principle working as Japanese people on behalf of the Japanese emperor. Industrial schools evolved into industrial workshops supplying military needs. And as the flow of Japanese industrial workers to the colony slowed and reversed, local workers took advantage of the gaps left behind. Ambitious manual laborers became assistant technicians, assistant technicians replaced technicians, and technicians climbed up to claim the elevated identity of engineer in the workplace. In 1940, one Tainan graduate became the first local employee in the Governor-General's railway bureau

Fig. 2.8 Mechanical engineering students in the forging workshop at Taipei Industrial School, 1942 (Courtesy of National Taipei University of Technology)

while another gained a senior teaching position at his alma mater (Kao 2011, 50–57). For his part, Chu Jiang-Huai rose to become professional engineer for Taiwan Electric Company in 1941 (Chu 2003, 90). And by the late 1930s, the number of local companies had grown so fast that they had become the majority in every industry (Hsieh 2013).

Heavy U.S. bombardment of the island during the war destroyed most of its large factories. Japanese industry ground to a halt as between 60% and 85% of power plants were completely destroyed, and nearly 50% of major industrial assets along with several significant transportation and communication infrastructures were heavily damaged (C.y. Tu 2017, Zheng et al. 1991, 216, 226, Wu 2023). But since no invasion of the island took place, the small and medium-sized local companies remained able to operate to the extent they were able to acquire raw materials and hang onto customers.

In the many gaps that appeared, local companies flourished. The Xie-Zhi Business, for example, rose from its status as a sub-contractor to an active bidder on construction jobs across the island. It became the main contractor for a new airport in Taipei and for the early-stage construction of airports in Hsinchu, Taichung, and Tainan. Its managers had all received their education at Taipei Industrial School. During the war, the Governor-General's office managed the quality control of local companies by limiting the founding of new firms and requiring the integration of existing firms. The owner of Xie-Shi, Lin Shih-Tsao, allowed former employees and business partners to establish companies under the auspices of his company. These were subsidiaries in name only. Also, in parallel with other companies, Xie-Zhi integrated its business vertically. During a construction project, the principal contractor needs skilled labor, architects, supervisors, accountants, etc. as well as raw materials such as sand and steel. Xie-Zhi took advantage of new freedoms to expand into related industries, such as metal-working and forestry. The Guang-Zhi Business, a construction company, did the same (Cheng 2012, 82–92).

At the end of the war, Taiwan was still part of the empire but had also become something outside of it. During the colonial period, some local men who had gained industrial knowledge and practices were able to freelance around Japanese industry. By 1945, industrial technicians and would-be engineers increasingly contributed not only around it but also within it.

References

Chang, Tsung-Han 張宗漢. 1980. *Guang fu qian tai wan zhi gong ye hua* 光復前台灣之工業化 *(Taiwan's Industrialization before Retrocession)*. Taipei: Linking Publishing (聯經).

Chen, Chiu-Kun 陳秋坤, and Cheng-Tian Kuo 郭承天. 1991. "Wo yu dong yuan dian ji gong si: Lin chang cheng xian sheng fang wen ji lu 我與東元電機公司: 林長城先生訪問紀錄 (Me and TECO: Lin Chang-Chen's Interview)." *Kou shu li shi* 口述歷史 *(Oral History)* 2: 185–206.

Cheng, Li-Ling 鄭麗玲. 2009. "Jin dai hua, ping deng hua yu cha bie hua zhi jian: Tai bei gong ye xue xiao xue sheng zhi jiu xue yu jiu ye 近代化、平等化與差別化之間：臺北工業學校學生之就學與就業 (1923~1945) (Modernization, Equalization and Differentiation: Education and Employment of Students in Technical Schools of Taipei (1923–1945))." *Tai wan shi yan jiu* 臺灣史研究 *(Taiwan Historical Research)* 16(4): 81–114.

Cheng, Li-Ling 鄭麗玲. 2012. *Tai wan di yi suo gong ye xue xiao: Cong tai bei gong ye xue xiao dao tai bei gong zhuan (1912–1968)* 台灣第一所工業學校: 從台北工業學校到台北工專 *(The First Indusrial School in Taiwan: From Taipei Industrial School to Taipei Technical College)*. New Taipei City: Daw-Shiang (稻香).

Cheng, Li-Ling 鄭麗玲, and Li-Chu Yang 楊麗祝. 2009. *Tai bei gong ye sheng de hui yi* 臺北工業生的回憶 *(Former Students' Memory of Taipei Industrial School)*. Vol. 1. Taipei: National Taipei University of Technology.

Zheng, Youkui 郑友揆, Cheng Linsun 程麟荪, and Chuanhong Zhang 张传洪. 1991. *Jiu zhong guo de zi yuan wei yuan hui (1932–1949): Shi shi yu ping jia* 旧中国的资源委员会 *(1932–1949):* 史实与评价 *(National Resources Commission in Republican China, 1932–1949: Historical Facts and Evaluation)*. Shanghai: Shanghai Academy of Social Sciences Press.

Ching, Leo T. S. 2001. *Becoming Japanese: Colonial Taiwan and the Politics of Identity Formation*. Berkeley, CA: University of California Press.

Chu, Jiang-Huai 朱江淮. 1980a. "Hui gu tai wan sheng ke xue zhen xing hui chuang li wu shi zhou nian 回顧臺灣省科學振興會創立五十週年 (Review of FAAS's Past 50 Years)." *Tai wan ke xue* 臺灣科學 *(Formosan Science)* 34 (3–4): 5–7.

Chu, Jiang-Huai 朱江淮. 1980b. "Tu tsung ming mi shou zhu ci yu jin bao shi 杜聰明米壽祝詞 1980.8.24於金寶石酒樓 (Celebrating Tu Tsung-ming's Birthday Speech at Jinbaoshi Restaurant on Aug. 24, 1980)." *Tai wan ke xue* 臺灣科學 *(Formosan Science)* 34 (3–4): 3–4.

Chu, Jiang-Huai 朱江淮. 2003. *Zhu jiang huai hui yi lu: Tai ji di yi wei dian qi gong cheng shi* 朱江淮回憶錄: 台籍第一位電氣工程師 *(Memoirs of Chu Jiang-Huai: The First Taiwanese Electrical Engineer)*. Taipei: Chu Jiang-Huai Foundation (朱江淮基金會).

Downey, Gary Lee, and Masanori Wada. 2011. "Avoiding Inferiority: Global Engineering Education across Japan." American Society for Engineering Education Conference & Exposition, Vancouver.

Han, Kyonghee 한경희. 2021. *The Making of Korean Engineers: Historical Formation and Development* 한국 엔지니어의 형성과 발전. Paju: Dulnyeok (들녘).

Hsieh, Kuo-hsing 謝國興. 2013. "Zhan hou chu qi taiwan zhong xiao qi ye de zhi min di chuan cheng 戰後初期臺灣中小企業的殖民地傳承 (Taiwan Small and Midium Companies' History during Colonial Period and Its Impacts after World War Two)." History of Border Area and Their Shaping of Subjectivity: Proceedings of the 4th Sinology Conference, Taipei.

Hsu, Hsueh-chi 許雪姬, ed. 2002. *Ri zhi shi qi zai man zhou de Tai wan ren* 日治時期在滿州的臺灣人 *(Taiwanese in Manchuria during the Japanese Occupation)*. Taipei: Institute of Modern History, Academia Sinica.

Hsu, Pei-hsien 許佩賢. 2013. "Ri zhi shi qi de tai wan bu xi xue xiao 日治時期的臺灣補習學校 (Vocational Supplementary Schools in Taiwan during the Japanese Occupation)." *Shi da tai wan shi xue bao* 師大臺灣史學報 *(Bulletin of Taiwan Historical Research, NTNU)* 6: 101–148.

Hsu, Pei-hsien 許佩賢. 2014. "Ri zhi zhong qi de gong xue xiao bi ye sheng yu tai wan she hui 日治中期的公學校畢業生與臺灣社會 (Common School Graduates and Taiwanese Society during the Middle Period of the Japanese Occupation)." *Guo shi guan guan kan* 國史館館刊 *(Bulletin of Academia Historica)* 41: 133–156.

Hsu, Sheng-Kai 徐聖凱. 2009. "Ri zhi shi qi tai bei gao deng xue xiao zhi yan jiu 日治時期台北高等學校之研究 (A Study on Taipei High School during the Japanese Colonial Period)." Master Thesis, History, National Taiwan Normal University.

Huang, Shih-chuan 黃士娟. 2012. *Jian zhu ji shu guan liao yu zhi min di jing ying 1895–1922* 建築技術官僚與殖民地經營 *1895–1922 (Building Technocrats and Colonial Enterprise, 1895–1922)*. Taipei: Yuan-Liou (遠流).

Kao, Shu-Yuan 高淑媛. 2002. "Ri zhi shi qi tai nan gao deng gong ye xue xiao he tai wan gong ye hua 日治時期臺南高等工業學校和臺灣工業化 (Tainan Junior College of Engineering and Taiwan's Industrialization During the Colonial Period)." In *Cheng gong de dao lu: Di yi jie cheng gong da xue xiao shi xue shu yan tao hui lun wen ji (The Road to Success: Conference Proceeding of NCKU History)* (成功的道路: 第一屆成功大學校史學術研討會論文集*)*, edited by Department of History at NCKU. Tainan: National Cheng Kung University.

Kao, Shu-Yuan 高淑媛. 2011. *Tou leng xiong kuan jiao min: Cheng da zao qi bi ye sheng yu tai wan gong ye hua* 頭冷胸寬腳敏: 成大早期畢業生與台灣工業化 *(NCKU's Early Graduates and Taiwan's Industrialization)*. Tainan: National Cheng Kung University.

Kao, Shu-Yuan 高淑媛. 2016. *Tai wan gong ye shi* 臺灣工業史 *(Taiwan Industrial History)*. Taipei: Wunan (五南).

Komagome, Takeshi 駒込武. 2019. *Tai wan ren de xue xiao zhi meng: Cong shi jie shi de shi jiao* 「臺灣人的學校」之夢：從世界史的視角看日本的臺灣殖民統治 *(Taiwanese School Dreams: From the Perspective of World History)*. Translated by Shou-pin Su 蘇碩斌, Pei-hsien Hsu 許佩賢 and Shi-ting Lin 林詩庭. Taipei: National Taiwan University.

Kuo, Wen-Hua 郭文華. 1996. "Tai wan yang wu ke ji chu tan (1887–1995): Cong tai wan tie lu, tai bei ji qi ju yu ji long mei kuang chu fa de chu bu tao lun 臺灣洋務科技初探(1887–1895): 從臺灣鐵路、台北機器局與基隆煤礦出發的初步討論 (Technology in Taiwan's Self-Strengthening Movement (1887–1895): A Preliminary Discussion of the Taipei Machinery Bureau, Taiwan Railway and Keelung Coal Mine)." *Xin shi xue* 新史學 *(New History)* 7(2): 99–138.

Li, Yuan-Hui 李園會. 2005. *Ri ju shi qi tai wan jiao yu shi* 日據時期臺灣教育史 *(The History of Education of Taiwan under Japanese Rule)*. Tainan: Fuwen (復文).

Liao, Wei-Cheng 廖偉程. 1993. "Ri ju tai wan zhi min fa zhan zhong di gong chang gong ren (1905–1943) 日據臺灣殖民發展中的工場工人 (1905–1943) (The Workers in Colonial Taiwan (1905–1943))." Master Thesis, History, National Tsing Hua University.

Lin, Chung-Shen 林忠勝. 2007. *Liao qin fu hui yi lu* 廖欽福回憶錄 *(Liao Ching-fu Memoir)*. Taipei: Avangard (前衛).

Lin, Heng-tao 林衡道. 1977. *Tai wan Shi* 臺灣史 *(History of Taiwan)*. Nantou, Taiwan: Historical Research Committee of Taiwan Province.

Lin, Ting-Sheng 林挺生. 1951. "Tan tai wan gong ye 談臺灣工業 (On Taiwan's Industry)." *Tai wan ke xue* 臺灣科學 *(Formosan Science)* 5 (1–2): 8–9.

Liu, Yi-Chang 劉益昌, and Chu-Ching Lin 林祝菁. 2008. *Lin ting sheng chuan: Jiao shou, xiao zhang, dong shi chang* 林挺生傳: 教授, 校長, 董事長 *(Biography of Lin Tingsheng: Professor, Principal, Chairman)*. Taipei: CTU Publications (資訊文化).

Myers, Ramon Hawley, and Mark R. Peattie, eds. 1984. *The Japanese Colonial Empire, 1895–1945*. Princeton, NJ: Princeton University Press.

Preparatory Working Group for Professor Lai Tsai-te Memorial Academic Symposium, 賴再德教授紀念學術研討會籌備工作小組, ed. 1997. *Zhui si: Lai tsai te jiao shou ji nian wen ji* 追思: 賴再德教授紀念文集 *(In Memoriam: Memorial Collected Works of Professor Lai Tsai-te)*. Tainan: Department of Chemistry Engineering, National Cheng Kung University.

Tainan Junior College of Engineering, Taiwan Director-General. 1934. *Tai nan gao deng gong ye xue xiao yi lan* 台南高等工業學校一覽 *(Tainan Junior College of Engineering Guildline)*. Tainan: Tainan Junior College of Engineering.

Taipei Industrial School. 1935. *Tai bei gong ye xue xiao xue ze* 台北工業學校學則 *(Regulations of Taipei Industrial School)*. Taipei: Taipei Industrial School.

The Department of Electrical Engineering at National Taiwan University. 2016. *Guo li tai wan da xue dian ji gong cheng xue xi xi shi: 1945–2015* 國立臺灣大學電機工程學系系史*: 1945–2015 (The History of the Department of Electrical Engineering at NTU: 1945–2015)*. Taipei: The Department of Electrical Engineering at National Taiwan University.

Tsurumi, E. Patricia. 1977. *Japanese Colonial Education in Taiwan, 1895–1945*. Cambridge, MA: Harvard University Press.

Tu, Chao-Yen 涂照彥. 1993. *Ri ben di guo zhu yi xia de tai wan* 日本帝國主義下的台灣 *(Taiwan under Japanese Imperialism)*. Translated by Ming-Chun Li 李明峻. Taipei: Ren Jian (人間).

Tu, Cheng-yu 杜正宇. 2017. "Lun er zhan shi qi de tai wan da kong xi, 1938–1945 論二戰時期的臺灣大空襲 (1938–1945) (The Bombing Campaign on Formosa during World War II, 1938–1945)." *Guo shi guan guan kan*國史館館刊 *(Bulletin of Academia Historica)* 51: 59–95.

Tu, Tsung-ming 杜聰明. 1970. "Zhui dao ben hui chuang hui hui yuan wang chao ying xian sheng 追悼本會創會會員王超英先生 (In Memory of Our Founding Member Mr. Wang Chao-Ing)." *Tai wan ke xue* 臺灣科學 *(Formosan Science)* 24(3–4): 55.

Wang, Yao-te 王耀德. 2012. "Ri zhi shi qi tai nan gao deng gong ye xue xiao zhi chou ban yu jian xiao 日治時期臺南高等工業學校之籌辦與建校 (The Organizing and Founding Process of Tainan Technical College during the Japanese Colonial Era)." *Tai wan wen xian* 臺灣文獻季刊 *(Taiwan Literature)* 63(1): 21–56.

Wang, Yao-te 王耀德. 2011. "Ri zhi shi qi tainan gao deng gong ye xue xiao zhi ru xue wen t iyu zu qun 日治時期臺南高等工業學校之入學問題與族群關係 (Founding of Tainan Technical College during Japanese Colonial Era)." *Tai wan shi yan jiu* 臺灣史研究 *(Taiwan Historical Research)* 16(2): 23–62.

Wu, Tsong-min 吳聰敏. 2023. *Tai wan jing ji si bai nian* 台灣經濟四百年 *(A 400-Year History of Taiwanese Economy)*. Taipei: Spring Hill (春山).

Yanaihara, Tadao 矢内原忠雄. 1999. *Ri ben di guo zhu yi xia zhi tai wan* 日本帝國主義下之台灣 *(Taiwan under Japanese Colonialism)*. Translated by Hsien-Wen Chou 周憲文. Taipei: Haixia Xueshu (海峽學術).

3 Climbing Mainland Infrastructures

The moment promised a bright future for engineers in Taiwan, just not necessarily engineers who were born there. In early 1956, slightly more than a decade after the war against Japan, an elite collection of government ministers in Taiwan gathered to attend the 60th anniversary of an engineering school. During the celebration, they agreed to re-start it.

The engineering school had originally been founded across the strait in 1896 as Nanyang Public School and then evolved to become Jiao Tong University of Science, Engineering and Management in Shanghai in the 1930s (Jesiek and Shen 2012). During its roughly two decades of rule, the government of the Republic of China had relied heavily on graduates from the school to lead what the governing nationalist party, the Kuomintang (KMT), called "national reconstruction" (Chiang and Jaffe 1947, Y.S. Sun 1920). When the KMT government fled the mainland for Taiwan in 1949, it left the school behind but continued its policy of national reconstruction through government-led science, education, and production. By the mid-1950s, the government was ready to add research, and restarting the school as Chiao Tung University could help lead the way. The transplanted school resumed modestly with a small research institute and graduate program in electronics. The program's faculty consisted entirely of mainlanders (Y. Wu 2016a, b).

This chapter follows the transfer of infrastructures of engineering education and work from mainland China to Taiwan after what the KMT government called the War of Resistance. It examines their subsequent developments and the efforts of local engineers and would-be local engineers to join and climb them. As the war was ending, enterprising local engineers and small business owners planned to continue advancing industry for Taiwan as a new province of the Republic of China. However, they soon encountered hierarchical infrastructures of education and work that championed and privileged forms

K.-H. Chang et al., *Engineers and the Two Taiwans*, Synthesis Lectures on Global Engineering, https://doi.org/10.1007/978-3-031-59766-4_3

of knowledge they did not possess. Once again, ambitious industrial workers and would-be engineers would find themselves starting on the underside, looking up. And once again, they would look for and find ways to climb, both within and around those hierarchies.

Arriving Ethical Knowers for National Reconstruction

After the Japanese military officially handed over Taiwan to the Republic of China on October, 25, 1945, the republican government moved quickly to take ownership of private Japanese companies and begin consolidating them. First, it merged 60 of the largest into large state-owned companies identified by province and in basic industries, e.g., Taiwan Sugar, Taiwan Power, Taiwan Aluminum, and Taiwan Cement. To operate them, it dispatched managers from its massive National Resources Commission (國家資源委員會, 1932–1952) (Hong 2016). Second, the republican government combined more than 300 other companies into 18 new ones and assigned them to the newly-established provincial government on the island. In this case, commission officials and provincial officials would manage the provincial companies together. Third, the government granted authority over another 92 Japanese companies to newly-established county authorities within the province to serve more local needs. And fourth, the ruling Kuomintang party gave itself title to a final set of 19 companies (Japanese Asset Management Committee of the Taiwan Province Takeover Commission 1947).

By taking direct ownership of companies and assigning commission officials to manage them, the republican government introduced a longstanding practice from the mainland that blurred the boundary between public and private industry and bore a responsibility to advance the official project of national reconstruction. More than two decades earlier, the founding leader of the republic, Sun Yat-Sen (1866–1925), had theorized the "reconstruction" of China to prevent it from becoming a "dumping ground' for the... overproduction [of] commercial nations" following the end of World War I (1914–1918). China, he asserted, needed to "adopt... machinery" and "nationaliz[e]... production" to achieve genuine "unification" (Y.S. Sun 1920, ii). In what became known as his "Industry Plan," Sun outlined a massive program for constructing ports, railways, canals, and productive industries across China, as well as for sharpening the boundaries around it as a republic. "[M]y idea," Sun wrote, is "to make capitalism serve socialism in China...." That is, the "industrial development of China should be carried out... by private enterprise and... national undertaking" (Y.S. Sun 1920, 165).

At the time, it was not a category mistake to imagine private enterprise and national undertaking joining together to serve the common good of the Chinese people. Writing shortly after the Russian revolution (1917) and the so-called war-to-end-all-wars (World War I), Sun contested the capitalist image of struggle in economic relations at a time when socialist visions were attracting unprecedented support across the northern hemisphere. "[I]t has been discovered by post-Darwin philosophers," Sun asserted without evidence, "that the primary force of human evolution is cooperation and not struggle as in the animal world" (Y.S. Sun 1920, 160). Individual economic interests and activity were, in his view, not necessarily barriers to cooperative action on behalf of a collective whole.

Yet because of protracted conflict across the mainland following the fall of the Qing dynasty in 1911, achieving a fully-reconstructed China basically meant constructing a bounded country for the first time. Private enterprise would serve the common good in this country, according to Sun, because it provided needed benefits to the Chinese people. He did not elaborate exactly how private initiatives would cooperate with one another, but he did define national undertaking as filling in where private enterprise failed to serve. In China, national undertaking would also require international cooperation, for as a predominantly rural country it lacked significant investment capital. Completing a multitude of construction projects would require hefty funding through, hopefully, assistance from other countries. Sun published his industry plan in English to help attract foreign investors, especially from the United States.

When Sun's party, the Kuomintang, gained political control of China in 1928 under the leadership of Chiang Kai-Shek (1887–1975), the new nationalist government quickly established the Construction Commission of the National Government (國民政府建設委員會, 1928–1938). Quite explicitly in the "spirit" of Sun Yat-Sen's plan, the commission was charged to "research, prepare and complete a Reconstruction Plan for the whole country" (Kirby 2000, 141). The commission's 39 members included all cabinet ministers, all heads of bureaus charged to lead reconstruction in the provinces, and the mayors of the republic's two major cities, Shanghai and Nanjing. It represented essentially every sector of government.

The Construction Commission became the government's main vehicle for national reconstruction. It established and managed large public companies, reaching a total of 130 by 1945. It sought to maximize cooperative initiative in private enterprise by building a huge regulatory apparatus that implemented technical standards, regulated prices, and limited profit margins to 25% percent of invested capital. By 1945, the commission had grown to become a massive industrial bureaucracy, employing nearly 64,000 officials (Kirby 2000, 142–143, 147 (cit. n. 24); Zheng et al. 1991, 107–113).

Along the way, it also gained a more circumscribed name. Sun had outlined an economic vision without a plan for its implementation or theorizing its precise relation to his more general vision of "revolutionary reconstruction." Chiang Kai-Shek filled in the gaps by formulating specific plans for education, the economy, and the military. National reconstruction, Chiang wrote in 1943, must consist of "five main lines of reconstruction—psychological, ethical, social, political, and economic" (Chiang and Jaffe 1947, 161) Psychological reconstruction would "reviv[e] … the nation's ancient culture" and link it to the "cultivation of genuinely scientific knowledge." Ethical reconstruction would "fulfill the principle of complete loyalty to the state and of filial piety toward the nation." Social reconstruction would revive the notion that "the rural districts and villages constitute the foundation of the state." Political reconstruction included "develop[ing] democratic institutions and the building of a strong system of national defense," most importantly to resist Japanese imperialism. Lastly, in economic reconstruction, "particular emphasis must be given to industrialization" (Chiang and Jaffe 1947, 163, 165, 167, 169, 171). In this elaborated framework, the technically-oriented commission could no longer claim to be managing national reconstruction as a whole. The government delegated its works to the National Resources Commission.

During the first four years of Chinese republican rule in Taiwan, which the first governor-general called the "Retrocession,"[1] the National Resources Commission dispatched to the province more than a thousand officials who held university degrees in engineering (W. Chu 2017, 192). The most elite of these officials had completed their degrees at Jiao Tong University in Shanghai, bringing with them the identity of scientific specialists. That is, they had qualified to serve as technical "scholar-officials" in technical bureaucracies without political affiliation (Chang et al. 2020, 453–456). They moved quickly into management positions at the newly-acquired industrial enterprises and within the provincial government itself. Their engineering knowledge held special status. It qualified them as moral leaders.

Qing dynasty officials had formally elevated the status of scientific knowledge during the self-strengthening movement. As described in Chap. 1, the infrastructure of classical learning had for centuries produced scholars who gained access to government posts through mastery of ancient texts and superior performance on exams (Kirby 2010, 305). They became officials who understood the privileges and responsibilities of ethical leadership in neo-Confucian terms. The evolving Qing court affirmed that science education also warranted the status of higher learning by adding a new civil service exam that tested competencies in scientific knowledge in 1905 (Guan 2008). And it founded two tertiary-level schools, including the one that would become Jiao Tong University, to guide students in rigorous study and preparation.[2]

Construing science knowledge as, in part, moral knowledge shifted the focus away from valuing scientific knowledge for its own sake, emphasized in Anglo-American contexts, and toward defining and assessing that knowledge in terms of its potential material benefits. In republican China, the schools' curricula aimed to empower the republic by providing students with access to usable scientific knowledge and career pathways that would advance industrial and, hopefully, military productivity. The key word was "productivity." In the face of obvious military power possessed by both Western countries and Japan, the "material scientism" in Chinese science came to "mean technology and productivity, which are not included in the definition of science in the West." (Jin and Liu 2004, 149). Moral leadership came to include scientific leadership as evidenced by China's military and industrial strength in relation to adversaries who threatened it.

[1] The term "retrocession," meaning ceding back, became contested in later decades since the receiving entity was the Republic of China, different from the Qing dynasty that had in 1895 ceded the province of Taiwan to the Japanese empire.

[2] The other became Tianjin University, located in a coastal trading city downstream of Peking.

Sun Yat-Sen's Industry Plan had explicitly attached practices of science to the project of national reconstruction, highlighting material benefits. He imagined cultivated land being improved by "scientific methods" and "modern machinery." He called for a "scientific bureau" to guide a massive expansion in the industrialized production of silk. "[S]cientific methods," he argued, would also "improve the quality" and "increase the quantity" of the production of wool. And greatly expanding productive industries would require extensive, country-wide infrastructures for transportation and communication. Sun's vision pictured China as recovering its position in the world as the center of human civilization, but only by making sure that national reconstruction was properly "based on knowledge, that is upon scientific planning" (Y.S. Sun 1920, 138, 144, 145,147).

It was in Chiang Kai-Shek's hands that engineering emerged as a key form of knowledge for realizing science as moral enterprise via production for the whole. After his party gained control of government in 1928, Chiang worked to build educational infrastructures to cultivate knowledge for production. The "development of education," in his framing, was a "fundamental task" in national reconstruction, alongside military affairs and economic development. (Chiang and Jaffe 1947, 160). And education for production had to become more than higher scientific learning.

With an eye toward the United States, Chiang championed "vocational education" at the secondary level. Students would gain introductions to scientific learning and then acquire approaches to implementing it via the practices of work. The United States Congress had itself formalized the term on the eve of entering World War I in 1917. Involving the application of science to more than industry alone, vocational education included curricula in commerce, home economics, agriculture, and the trades. Military readiness, it was argued, required adequate preparation across a range of vocations (Carleton 2002, 70–74). In China, the emphasis became industry. The nationalist government under Chiang issued in 1932 the Vocational School Rule (職業學校規程) to build robust infrastructures of learning at the primary and secondary levels for productive work, meaning work to develop industry. The rule required that by 1937 the budget for vocational schools must not be less than 35% of the total budget for secondary schools (Education Yearbook Compilation Committee of the Ministry of Education 1948).

Construing productivity in national reconstruction significantly as industrial productivity had the effect of identifying higher education in science increasingly with degrees in engineering. A degree in engineering came to mark a supreme achievement in learning that promised to advance industrial production as a morally benevolent end, especially important during wartime. Students followed. By the end of the War of Resistance, the number of university graduates completing degrees in humanistic disciplines had dropped below 10%, while the number in the natural sciences approached 10% and the number in engineering exceeded 18 percent (Kirby 2010, 305; K. Chang 2013, 97).

After the war, Chiang published a detailed policy plan for the next decade of national reconstruction, emphasizing practices for education, the economy, and the military. In Chiang's plan, the number of graduates pursuing vocational careers must increase by at least a factor of five. The increase must include at least half a million graduates in "technical education from vocational middle school and high school" and at least a half million "university and technical school graduates" the majority of whom would earn degrees in engineering. Through graduates such as these, he argued, China could not only "preserve the inherited ethics of the nation and restore its traditional wisdom and ability," it could also "master the advanced science of the West and introduce the newest Western technical skills." Above all, through production-oriented learning and expanded production, Chinese people could decisively re-affirm their "unwillingness to become slaves" to Japanese imperialism (Chiang and Jaffe 1947, 176–77, 240, 97).

The engineering managers who relocated to Taiwan beginning in 1945 arrived as elite knowers for national reconstruction. Their university degrees marked their knowledge as far superior to anything held by the vast majority of provincial workers, managers, and, even, industry owners (Y. Yen 1989). The official government line was that the Japanese colonialists had treated the population on Taiwan as slaves, providing them with only a minimum of "enslaved education." A 1945 report from the Ministry of Education, for example, lamented that "vocational education in Taiwan" at the lowest levels had been "overly developed." It was clear the Japanese rulers had used vocational schools as "instruments to enslave compatriots in Taiwan" (T. Chen 2002; Ou 1997, 46). A policy maker from the ministry elaborated, writing that an overwhelming number of Taiwanese under Japanese rule had been prohibited from attending middle schools and preparing for higher education. The "low-level education" that built Japanese identity and loyalty to the Japanese emperor and limited graduates to lower-ranked jobs, he asserted, had restricted the Taiwanese people as a whole (Hsueh 1990).

Yet there was also anxiety among the arriving leadership about how its plans might be received. A decade earlier, the party official who would become the first governor-general, Chen Yi (陳儀;1883–1950), had represented Chiang Kai-Shek at an exposition in Taiwan organized by the Japanese government to commemorate its forty years of "administration." Chen's report on the visit had praised the strong industrial development on the island, its advanced public services, and the elevated quality of life experienced by the people there (Han 2016). Another decade would likely have brought further improvements, difficult to match. In addition, there was much uncertainty about what it would take to win the loyalty of the most elite Taiwanese leaders who had been educated in Japan and were familiar with their practices. As Chen actually faced the prospect of taking over Taiwan, he wrote to the minister of education that they must "show to Taiwanese people the efficacy of the revolution, and not become a joke to the enemy." He wanted 10,000 new teachers as soon as possible to bring Taiwan fully into the project of national reconstruction (J. Chang 1990, 53–55).

Not Accepted as Knowers for Reconstruction

Fewer than three weeks after what the KMT government called "Retrocession Day," a collection of industrialists on the island of Taiwan hurriedly formed the Taiwan Institute of Industrial Technologists (台灣工業技術人會). Their stated goal was to come together "to serve the nation with industry" (Anonymous 1945a). To be able to participate in resuming industrial development on the island, hopefully at managerial levels, the all-male industrialists had to make themselves visible and their knowledge legible to arriving officials from the National Resources Commission (Anonymous 1945b). They claimed the identity of technologists to elevate their status among officials who held engineering degrees. U.S. bombing during the war had destroyed many factories. Nearly all Japanese engineers and technicians had departed, and industrial production of military and consumer goods had largely ground to a halt (T. Chang 1980, 62). Perhaps the imperial glass ceiling that had held them down for decades could now be lifted. The Taiwan Institute of Industrial Technologists quickly received 1,200 applications for membership.

The industrialists also sought to distinguish themselves amidst a broader movement of male technical experts who had announced themselves to the new government as usable science professionals. Many had recently joined with physicians, teachers, and technical specialists in agriculture and forestry to re-brand and expand the small Taiwan Polytechnic Society into the mass Formosan Association for the Advancement of Science (臺灣科學振興會) (Anonymous 1945c). This move conveyed an argument for visibility and plea for participation. Within the colonial regime, the term "polytechnic" had labeled technical practices to support industry that were informed by but not driven or led by emergent engineering sciences. The new name championed a commitment to science that members knew the republican leadership embraced and valued. Its challenge was to squeeze its members into a definition of usability for national reconstruction.

As an association, members of the organization asserted that they were by no means "degraded people" who had fallen behind and below Chinese civilization, as one arriving general put it. Nor were they collaborators with Japanese rulers even though all could speak and write in Japanese. Adding the geographical marker "Formosa" in its English name also foregrounded its attachment to EuroAmerican spaces in which individual science professionals could flourish. About 600 members of the Taiwan Agriculture Comrade Society, for example, migrated into the new science association, along with more than 400 members of the Taiwanese-in-Japan Polytechnic Society (留日理工學會) who joined after returning to Taiwan (Li 2011, 7–11). Within a few months, the association had more than 3,000 members.

At the same time, the Chinese version of the name, 臺灣省科學振興會, translatable as Taiwan Provincial Association for the Advancement of Science, signaled acceptance of inclusion in the Republic of China. Explaining its creation to a mixed readership in 1950, then-president Tu Tsung-ming (杜聰明, 1893–1986), the first Doctor of Medical Sciences in Taiwan, asserted patriotically that "we were so delighted and appreciative that we return to our mother country China from Japan." Yet the substantial war damage and departure of the Japanese grounded serious concerns about the current state and future of industry on the island. "[W]e worried all the industrial, electrical and transportation systems would [not] continue to run smoothly as in the colonial period," he continued, "since they were run by the Japanese at the time." Recognizing the value that arriving officials attributed to science, he affirmed that the new association would "bring more talents together to help the Republic of China (ROC) government" (Tu 1950, 1).

Association leaders actively reached out to intellectuals on the mainland, trying to build individual networks and seeking advice on how to position themselves in the project of national reconstruction. Tu Tsung-ming, who had himself been educated at Kyoto Imperial University and served as researcher and teacher in pharmacology at the former imperial university in Taipei, traveled to Shanghai and Peking to deliver presentations at various universities about Taiwan's developments in medical science. He met the renowned contemporary philosopher, Hu Shih (胡適) (1891–1962), a former student of the educational philosopher John Dewey at Columbia University, who championed use of the Chinese vernacular in his writings. On his return, Tu reported findings about the "science facilities" and "medical education" in what he called the "inner land" (Tu 1948). During the colonial period, the term "inner land" had referred to Japan, using the same characters in both Japanese and Chinese. By putting the term in the title to his report, Tu made clear his understanding and acceptance of the island's altered geographical identity and new political status as a province of the Republic of China. The Formosan Association was more than prepared to serve an arriving government of scholar officials and to help train new talents (C. Yen 1950b).

The talents that came to number more than 3,000 did make clear their interest in working for Taiwan specifically and in positions that were appropriate to their knowledge. The engineer Chu Jiang-Huai, for example, reported in later memoirs that Chinese officials and colleagues "believed that the Japanese took us talented Taiwanese as their assistants," doing "only... what they demanded." "We Taiwanese scientists and engineers," he continued, "needed to debunk this kind of rumor," for things had changed during the late colonial period. As a result, it was important to "invite more young and talent [sic] Taiwanese to help the ROC government take over Taiwan."[3] Overall, association members wanted to "help the ROC government to repair infrastructural systems such [as] power, transportation and industrial systems," "maintain our level of expertise," and "improve industrial productivity to the level back before the Retrocession" (J. Chu 2003, 620–621). Unfortunately, local "scientists" could not be accepted as knowers of national reconstruction. Having been colonized for fifty years, they could not have completed the educational formation necessary to claim the status of committed technical leaders. The Formosan Association was not a Chinese association.

[3]Note that Chu wrote this in 2003 by which time it had become accepted for local people to call themselves Taiwanese. Such was not the case in the early days of the Retrocession when the KMT government asserted that local people were citizens of the Republic of China, however subordinate.

Over the two previous decades, graduates of Chinese universities on the mainland had themselves formed many professional organizations. It had become common for Chinese engineers and other graduates to seek advanced education in the United States and other countries, which qualified them as individual professionals (Wang 1984; Chinese Institute of Engineers 1946). In China, however, professional status had come to warrant governmental positions as "ethical" technical leaders. After all, Chiang Kai-Shek had defined the ethical "line of reconstruction" as "complete loyalty to the state and... filial piety toward the nation." The elite Chinese Institute of Engineers, an honorary association, supported a range of projects to advance industrial production in the republic (Ling 1961).[4] The so-called Society of 31 was named for the 31 members who were sent to the United States for advanced education in 1942, year 31 of the Republic. Most became officials in the National Resources Commission (Y. Cheng 1996). It was no surprise, therefore, that nearly half of the members of the Chinese Association for Sugar Production Technology moved to the new province to serve in the government-owned Taiwan Sugar Company (Y. Cheng 2009; W. Chu 2017, 193).

Claiming to be professionals serving the whole, members of the Formosan Association for the Advancement of Science sought to separate themselves from their colonial past. In addition to regular formal meetings, the association began publishing a professional journal in 1947 called *Formosan Science*. It also successfully gained support from locally-owned industrial firms and local industry associations, signaling a connection between professional work and private enterprise for the welfare of the whole. The local firms included Taiyan (台陽) Company, Tatung Company, and Taiwan Chemical Company, and the local industry associations included Taiwan Industry Society, Taiwan Machinery Society, and Taiwan Electrical Industry Society (C. Yen 1950a, 42).

However, following the huge influx of experts from the mainland in 1949, the government effectively dissolved the local association in 1950. The government "asked [the] Formosan Association to hold an all-member meeting... to reorganize itself" under governmental supervision (Tu 1950, 1). The leaders, who called themselves "core members," made an effort to do so by asking its members to re-register. Yet only 237 did so (C. Yen 1950a, 42, Anonymous 1950a). President Tu tried to rally them by encouraging them to "participate substantially rather than focusing on the growth of membership," make the association a "platform for experts in different disciplines," continue "monthly speech events," and "continue to issue the *Formosan Science*" (Tu 1950, 1). None of these proposals would succeed. While four issues of Formosan Science were published in 1948, only two appeared in 1949 and one in 1950.

In the only issue of the journal in 1950, the editor Yen Chao Pang (顏朝邦; 1918–1995) expressed regret that arriving Chinese scholar-officials had not accepted those already in

[4]The Chinese Engineering Society (中華工程師會), which merged together three similar societies, was formed in 1912 in Peking, China. The Chinese Engineering Institute was founded in 1917 in Ithaca, New York. In 1930, they joined together in Nanking, China to become the Chinese Institute of Engineers (Ling 1961, 1–7).].

Taiwan. "Thinking of unity," he was "sad" that "intellectuals in science and technology were not able to unite together in the early Retrocession." He judged that "it brought bad influence" to Taiwan, for "[s]cientists and technologists should think of unity to serve the country" and local members of the association were willing to do so. But the collapse of the organization–its reduction to 200 members–was "a stunning fact" (C. Yen 1950a, 42).

Meanwhile, the Taiwan Institute of Industrial Technologists never really got off the ground as a province-wide organization. Guilds of company owners tended to be local organizations, and incoming engineers from the mainland tended to dominate provincial activities. For example, they quickly formed a Taiwan Branch of the elite Chinese Institute of Engineers, founded the professional journal *Taiwan Engineering* (He 1947, 18, Anonymous 1950b, 48), and invited some of the most elite engineers from the province to join. Those invited included engineers that held degrees from Japan, such as Chu Jiang-Huai and Yen Chao Pang, as well as some graduates of the Taipei and Tainan institutes. That none of them would ever achieve significant positions within the organization indicated just how much the gesture was tokenism for a subordinate province. Not surprisingly, membership was wholly denied to any industrial managers, owners, or workers who had not completed secondary school, regardless of what they had subsequently accomplished in their careers (Anonymous 1950b, 1, 12). The first president of the Taiwan branch was, of course, a degreed engineer from the inner land (Ling 1961, 41).

In 1950, one local member of the Taiwan branch did try to persuade leaders from the mainland to accept more members from the province by explaining differences between mainland and local engineers. It was true, he acknowledged, that local engineers usually had less formal training for the Japanese had denied them access to higher education. Also, language was certainly an issue, as many local engineers had been raised and educated to live and work exclusively in Japanese. Yet, in contrast with most arriving scholar-officials holding degrees in engineering, local engineers possessed enormous experience doing actual engineering work in actual industrial settings. He pointed to clothing to illustrate the point. Akin to engineers in the United States or Europe, mainland engineers tended to wear nice dark suits to indicate professional accomplishment and status. By contrast, local engineers dressed informally. Indeed, the more senior their status, the older were the clothes they wore. This difference in appearance did not, however, indicate a difference in the value of their industrial knowledge (Anonymous 1950b, 1, 12).

Appeals such as these were to no avail. In 1950, the prominent local engineer/industrialist Yen Chao Pang looked back over the colonial period and the five years since the Retrocession, finding striking similarities between the barriers in each. Of those locals who went to Japan or Japanese-held territory to study or do research, "they eagerly came back to Taiwan" after the war "to contribute to their profession" in visible ways (C. Yen 1950b, 5). For example, soon after the Republic of China took control, his colleague Shih Chiang-Nan (施江南), then-director of the Formosan Association, had taken a group to provincial government officials to offer their services. Yet their offer was denied, akin to many others. Group members were surprised to be told that all positions had been taken by arriving engineers from the mainland (Hu 1980, 9).[5] They were confronting, Yen said, "the same situation that they experienced in the colonial period." That is, although they had "appropriate expertise," they found no appropriate places to contribute their profession." Shunned by a KMT government that did not want them, all gravitated toward employment in locally-owned industries, even those who had previously worked "in state or provincial enterprises" under the Japanese (C. Yen 1950b, 5).

Degreed Engineers Flow into Government Infrastructures

Sun Yat-Sen's Industry Plan had asserted that effective national reconstruction could merge individual initiative and cooperative action to benefit the Chinese people as a whole. It could not foresee the demographic situation found on the island of Taiwan after the War of Resistance in which a new government claimed to be acting for the Chinese people as a whole while also treating the island as conquered territory and subordinating its inhabitants. Taiwan was a province in name only. During the KMT government's first five years in Taiwan, it was preoccupied with civil war against the Communist Party on the mainland. After taking control of formerly Japanese companies, it exploited local economic production on the island to support the mainland. Low-paid Chinese soldiers assigned to the provincial government in Taiwan were permitted to support themselves through looting. In February 1947, political tension became active protest, revolt, and demands for reform. The provincial government responded with arrests and executions, killing more than 10,000 local people. It initiated what would become a forty-year period of martial law during which the KMT government would imprison more than 100,000 people and that would become known as the White Terror. Led by the authoritarian Kuomintang Party, the government justified martial law as necessary to protect "Free China" from the worldwide threat of Communism. And in 1949, the national government joined the provincial government on the island of Taiwan, co-existing side by side.

Engineers on the mainland had to choose whether or not to relocate to Taiwan with the Kuomintang government. The National Resources Commission formally elected to stay, becoming the only ministerial organization in the nationalist government to do so. Leaders believed they could help China more on the mainland even if such meant working with the Chinese Communist Party (CCP) (W. Chu 2017, 198–203; T.H. Wu 1988, 63–141; Y.C. Sun 1988, 14–62). The technical and ethical commitments of engineers were to China as a whole, not to a governing political party. How the commission could possibly fulfill this responsibility without being present in the mainland was uncertain at best.

But many degreed engineers decided to go. One group of officials from the National Resources Commission who went found positions within the provincial government.

[5] The president of Taiwan Institute of Industrial Engineers, Lim Bo-Seng (林茂生; 1887–1947), was also denied participation. Shih Chiang-Nan lost his life in 1947 in the February 28 Incident.

Some went to the new Provincial Committee of Taiwan Production (台灣區生產管理委員會) where they assumed responsibility for running the confiscated Japanese enterprises (Yu 2009, 52–53). Many who had been part of the Heavy Equipment Division moved to the new Bureau of Engineering Service within the Ministry of Economic Affairs. Bureau officials typically went to work on construction projects building or rebuilding roads, harbors, dams, and other transportation and communications systems. Most of these were civil engineers educated at Jiao Tong University and accustomed to government-led work (Liao 2013, 131–138; Lu 2000). The bureau would continue for the next three decades to both contract and manage construction projects across the whole of Taiwan.

A second group of degreed engineers from the National Resources Commission continued work as government employees by transferring to the army. In Chiang Kai-Shek's administration, "political reconstruction" included building a strong system of national defense. After 1949, successful defense required not only protecting the island from invasion but also planning to reclaim the inner land by military force. Government-led economic development on the island thus included weapons production and the construction of military installations for both defensive and offensive purposes alongside the construction of transportation and communication infrastructures. During the 1950s and 1960s, Chiang promoted two degreed engineers from the army, Chiang Shao (江杓) and Yang Chi-Tseng (楊繼曾), to lead the Ministry of Economic Affairs (1955–1958 and 1958–1965) (Liu 2005, 68). Indeed, during the forty-year period of martial law, ten of thirteen ministers of economic affairs would be degreed engineers who had been educated on the mainland.

The army actually established its own construction company, the Ret-Ser Engineering Agency (榮民工程處), in order to convert retired servicemen into a labor force. In 1954, it led construction of what would become the Cross-Island Highway linking the central cities of Taichung on the west and Hualien on the east. More than 10,000 former soldiers worked on the project that covered 260 miles of mountainous terrain, requiring 80 tunnels and 130 bridges (Anonymous 1976). The government-owned firm that grew out of the project (RSEA Engineering Corporation) would become the largest in the republic and, later, one of the top 100 construction companies in the world.

By far the largest collection of engineers arriving from the mainland found managerial positions in large industrial companies, both government-owned and private. The KMT government maintained the Sun/Chiang vision of national reconstruction through centralized governmental control by providing large companies with monopoly control in their industrial sectors as well as support for planning, design and project management. Many joined those that the government had created from Japanese firms, especially Taiwan Sugar, Taiwan Power, Taiwan Aluminum, and Taiwan Cement. Other engineers traveled with large private companies, such as the Chinese Petroleum Corporation, that relocated to Taiwan with assurances that the KMT government would continue providing capital investment. In the 1950s, the petroleum corporation used government support to establish the wholly-owned engineering consulting firm, China Technical Consultants, Inc., to help it extend its reach in the rapidly-expanding petrochemical industry. The Ministry of Transportation followed later by establishing Chinese Engineering Consultants, Inc. to support construction projects, and Taiwan Power extended its reach in power generation through Sinotech Engineering Consultants Inc. (K. Chang 2011).

In every case, governmental control also meant control by mainlanders, often mainlander engineers. Every firm and government agency preferred to recruit mainlander engineers for managerial and leadership positions rather than engineers born in Taiwan (Y. Lin 2014; Y. Chen 2014; Ho 2016). And the many degreed engineers who fled to Taiwan asserted their elite status by bringing with them the Chinese Institute of Engineers. It quickly absorbed the recently-formed Taiwan branch (Wang 1984, 3) (Fig. 3.1).[6]

Bringing National Reconstruction to Education

While Chen Yi, the first governor in the province, did not get his 10,000 teachers, he did begin implementing the project of national reconstruction through educational infrastructures. By emphasizing scientific knowledge and introducing science learning, the new government sought to begin integrating the island into a China that valued the preparation of loyal technical leaders and workers. Highlighting the orientation and moral edification that came with scientific knowledge, Chen Yi wrote that "Learning science is not limited to science itself." Rather, akin to benefits of classical learning, "scientific attitudes and methodology can also be applied to how we work and behave" (Y. Chen 1945, 2). To him, the enslaved education that people on Taiwan had endured under the Japanese had ensured their subjugation. It demanded that they serve the enemy's military ambitions and prevented them from understanding and benefitting from the value of science.

Prior to the Retrocession, Chen Yi's initial plan for bringing national reconstruction to Taiwan envisioned getting rid of all Japanese curricula, Japanese teachers, and Japanese researchers from education institutions (Taiwan Historica 1995, 297–298). Anticipating victory in the war, he chaired a Taiwan Survey Commission to draft guidelines for a takeover plan. According to those guidelines, the restored Chinese government should ban the Japanese language in all textbooks, government documents, and newspapers. It should also lay off all Japanese personnel from positions in government and schools through the secondary level, including technical schools (Taiwan Historica 1995, 315–324).

[6]The arriving members of the Chinese institute of engineers did cease publication of *Engineering*, its journal on the mainland, in favor of *Taiwan Engineering*. But as the KMT government continued in the 1950s to assert itself as the government of China, the institute chose in 1961 to revive the mainland journal. Rather than publishing the 15th volume of *Taiwan Engineering*, it issued the 35th volume of *Engineering*. Local engineers remained a minority throughout the period of martial law. Not until the 1980s would one win the Engineering Medal, the institute's highest prize for "scientific" accomplishment. See Chap. 4 to understand the shifts that took place (Wang 1984, 122–124).

Fig. 3.1 Chinese Institute of Engineers, 1961 annual conference (Courtesy of The Central News Agency, Taiwan)

After the war, the new republican government did begin reframing educational infrastructures in mainland terms. Distinguishing between national and provincial institutions, it introduced a hierarchy of scientific education over vocational education, and included industrial education within vocational education as a form of applied science. As primary schools added instruction in science, low-level industrial secondary schools became vocational schools teaching applied science. The former industrial school in Taipei gained a promotion to the status of a junior college, above vocational schools but below higher education.[7] The industrial school in Tainan fared even better, becoming the Taiwan Provincial College of Engineering. It was one of three colonial schools elevated to the status of higher education, alongside the Taiwan Provincial College of Agriculture and the Taiwan Provincial Teachers College. The formerly imperial university in Taipei became National Taiwan University (Figs. 3.2 and 3.3).

[7] Other industrial schools in Hsinchu, Taichung, Changhua and Chiayi remained low-level (J. Wu 1946).

Fig. 3.2 A chemical engineering student working in the intern factory of Taiwan Provincial College of Engineering, 1947 (Courtesy of National Cheng Kung University)

Fig. 3.3 Taipei Imperial University (Courtesy of National Museum of Taiwan History)

Transforming industrial education into vocational applied science brought explicit attachments to science knowledge. Curricula at the new junior colleges and provincial engineering college began reducing the numbers of credits attributed to practice sessions in shop-work, calculation, and problem solving called "experiments and exercise," and to replace them with additional credits in the engineering sciences. One former critic of industrial education under the Japanese, Wu Jao-Keng (吳饒畊, 1900–1976), endorsed the upgrades in the 1946 article "Promoting Vocational Education" published in *People News*. Wu explained how he had considered industrial education under the Japanese as low-level colonial education, oppressive for the Taiwanese people who were forced to endure it. However, "[n]ow that Taiwanese compatriots have become owners of the big Republic of China," he asserted optimistically, "Taiwan should pay attention to vocational education." The key difference for him was that applying the sciences promised meaningful opportunities with technologies. Harking back to the Self-Strengthening Movement's dream of saving China through science, Wu wrote that his "Taiwanese compatriots" must understand vocational education as "founding the nation by science and saving the nation by technologies" (J. Wu 1946).[8]

But curricular changes actually came slowly. The new government had few resources to allocate to the new province. As early as 1944, Chen Yi had alerted the Minister of Education that "[a]fter we reclaim [Taiwan], we will have to maintain necessary facilities and keep them nonstop" (J. Chang 1990, 53–55). And by June 1945, the Ministry of Education had adjusted its stance regarding Japanese instructors in one respect. Courses such as history, geography, and the Japanese language would still be wholly banned, but "[We will] try to retain those [Japanese] teaching technical subjects" (Ou 1997, 46). Chen Yi built this concern to continue technical teaching into an executive order issued shortly after Retrocession Day. It stipulated that "[i]ndustry and business do not close, the administration works nonstop, and schools will not be cancelled" (工商不停頓、行政不中斷、學校不停課) (Taiwan Historica 1995, 428).

Retrospective accounts by former students of the junior college in Taipei, for example, offered vivid memories of Japanese textbooks, Japanese teachers, and Japanese-language technical courses during the early years of the Retrocession. Chang Ping-Tung (張炳桐), a graduate who subsequently taught at the Chiayi junior college in 1947, remembered that "teachers [teaching technical courses] used Japanese textbooks and taught in Japanese, and students learned in Japanese." Also, Li Tsung-Lien (李宗廉), who went on to a career in engineering, recalled that "teachers receiving education from [Mainland] Japan and the USA taught in Japanese and Chinese respectively." In addition, "students had to choose between the Japanese or Chinese language for professional courses [on mechanics and electrical engineering] depending on which languages they were more familiar with"

[8] Much more known by the name Wu Cho-Liu (吳濁流), Wu's novel *Orphan of Asia* explored how Taiwanese people struggled to search for national belongingness during Japanese rule. The kin-based metaphor "orphan of Asia" remains common among many Taiwanese, depicting a country that has no seat in United Nations (see Chap. 4) and few diplomatic relationships with other countries.

(Cheng and Yang 2009, 170). National Taiwan University, other junior colleges and the provincial college also retained some one- to three-year technical courses that were taught in Japanese (T. Cheng 2002, 170; H. Lin 1999, 73–74).

Enrollments in vocational schools and junior colleges increased significantly as local families sent their children to participate in science and productivity. Seeing new opportunities within nationalist reconstruction, some even shifted their trajectories from medicine to engineering. The total number of schools remained relatively stable as the roughly 117 industrial secondary and industrial continuation schools in place in 1944 became a collection of 77 vocational schools in 1950. An enrollment totaling 15,000 initially dropped to 10,000 with the departure of Japanese students but then increased to more than 33,000 by 1949 (Statistics Office of the Chief Executive's Office of Taiwan Province 1946; Education Yearbook Compilation Committee of the Ministry of Education 1957).

Chen Yi also mobilized the nationalist image of science for productivity to justify retaining teachers and researchers at National Taiwan University. "[A]cademic research," he wrote, meaning research to advance science, "is intrinsically intended for human welfare, and... has no national boundary...." Hence the Chinese government must acknowledge that "Japanese governance has benefited from science" and "experimental institutes [at] Taipei Imperial University have explicitly and implicitly helped the industry." Accordingly, "we should... appreciate the efforts of Japanese researchers", i.e., appropriate them, and he himself would "try [his] best to develop all kinds of scientific research in Taiwan" (Y. Chen 1945).[9]

But the respect accorded to Japanese teachers and researchers did not carry over to the locals they had trained. The provincial government repopulated professorships at National Taiwan University largely with faculty from mainland universities. The department of civil engineering, for example, retained four Japanese professors while hiring nine from the mainland. Six of those had been educated or employed at Chiao Tung University or one of the other two most prestigious schools, Tsinghua University in Peking or Peiyang University in Tianjin (Department of Civil Engineering 1995, 14). By 1953, only six locals had gained faculty status in the College of Engineering along with 33 mainlanders, and all of the six were positioned as low-level lecturers or assistants (College of Engineering 2002, 64). At the Provincial College of Engineering in Tainan, only four of its 58 professors in 1950 were locals from Taiwan (T. Cheng 2002, 187).

[9]At the Imperial Taipei University, many local students had become interested in studying engineering prior to the Retrocession, thinking that engineers would have many opportunities after the war. "Some of them," according to Yen Yen-Chen "even gave up studying medicine (the most prestigious and promising profession for the Taiwanese in the Japanese rule), such as Chou Shui-Tu (周隨土) transferred his major from medicine to chemical engineering" (Y. Yen 1989, 80).

Climbing from the Underside, Again

In 1951, the owner of the locally-owned construction firm Tatung Company, Lin Ting-Sheng (林挺生, 1919–2006), published a paper in *Taiwan Engineering* asking the ROC government to stop favoring companies that came from the mainland and begin giving equal opportunities to local firms (T. Lin 1951b, 71). Lin had been trained in chemistry under the Japanese at Taipei Imperial University, graduating in 1941. Iconic of the struggles of local engineers for visibility and acceptance, Lin's paper appeared in the journal but was not listed in its table of contents (T. Lin 1951a). Yet the paper included a thinly-viewed threat that would prove to be a new, localized approach to struggle from within: if the ROC government did not heed their request for equal treatment, he and fellow industry leaders were going to build their own organization, the Taiwan Provincial Industry Society (台灣省工業會), to collaborate and support one another.

Local engineers and would-be engineers were largely unable to gain employment in the large state-owned industrial enterprises, let alone in the provincial or national government. As a result, they turned their sights to the private sector. Many locals who had been trained in industrial schools during the Japanese era established small companies after the war making consumer goods (Lin 2009, 93). The companies multiplied quickly, especially after the population increased dramatically in 1949. Between 1940 and 1953, the number of registered factories on the island increased from around 9,000 to more than 12,000. Textile factories in particular grew by a factor of 10 while mechanical parts factories grew by a factor of three and chemical product factories by a factor of two. The dramatic growth in consumer product production translated into increasing demand for graduates trained in mechanical and chemical engineering at the junior colleges and provincial school (Figs. 3.4, 3.5 and 3.6).

This expansion in the private sector gained significant support from the United States through its foreign aid and the political pressure that came with it. In 1948, the U.S. government joined with the Kuomintang Party to establish the Council for United States Aid. Anxious to protect capitalism from the threat of Communism, U.S. officials were skeptical of the concentration of power and resources that constituted national reconstruction in China. Their priority was industrialization in the private sector, and they provided much-needed funds to back it up. The Aid Council became the major department supporting industrialization of Taiwan, including local companies (Chang et al. 2020; T.M. Wu 2023, 338–368, 395, 400–401).

Between 1950 and 1965, U.S. Aid to the Republic on Taiwan would total nearly $3.7 billion USD. By the late 1950s, funds from U.S. aid amounted to 12.5% of the KMT government's annual budget (T.M. Wu 2023, 341). Nearly half of those funds did support large transportation and communication projects, such as electricity generation and distribution, Taichung and Kaohsiung harbors, the telephone system, Shimen Reservoir, and the Central Cross-island Highway. Much of the rest supported agriculture and various

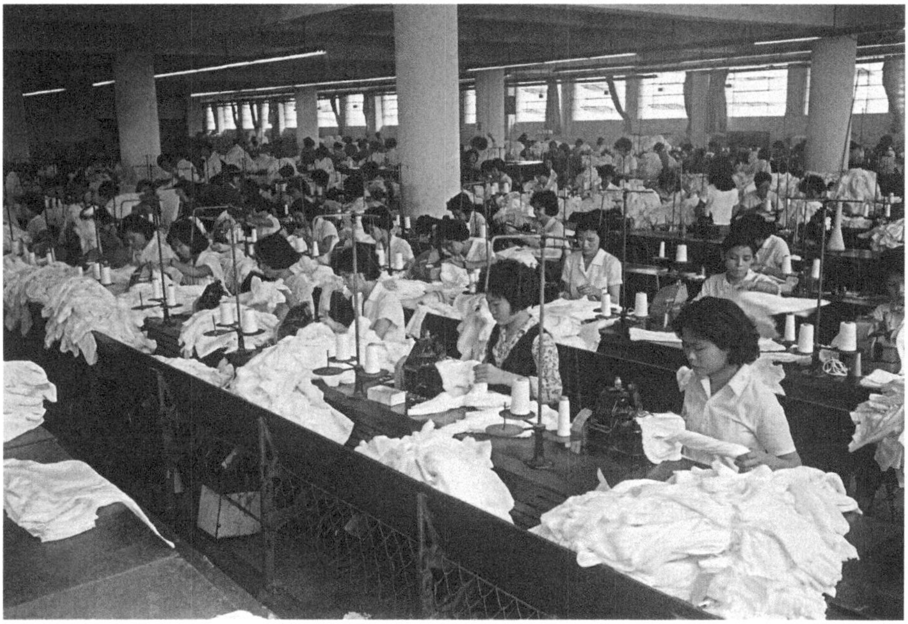

Fig. 3.4 Textile factory employees operating sewing machines, 1954 (Courtesy of National Archives Administration, Taiwan)

Fig. 3.5 Workers operating the plywood trimming machine in a local company, 1960 (Courtesy of National Archives Administration, Taiwan)

Fig. 3.6 Workers operating machines in a local company, 1971 (Courtesy of National Archives Administration, Taiwan)

industries. While these funds went primarily to large state-owned companies, small companies benefitted around the margins, just as they had done during the colonial period and the late Qing period before that (C. Lin 2012, 114–119; Kao 2016, 166–179) (Fig. 3.7).

While offering aid to support the KMT regime on Taiwan and enable its state-owned enterprises to work, the Americans also sought to promote a market economy on the island (Kao 2016, 238–239). A major initiative was the practice of land "reform," encouraging and pressuring rural landlords to interest themselves in industrial development by

Fig. 3.7 Cross-island highway (Courtesy of National Museum of Taiwan History and National Park Service, Taiwan)

converting their landholding into industrial holdings. Concrete measures taken in this connection were to transfer the state-and-province-jointly-owned Taiwan Cement Corporation and Taiwan Pulp and Paper Corporation, and the province-owned Taiwan Industrial and Mining Corporation and Taiwan Agricultural and Forestry Development Corporation to private ownership by offering the government stock for sale to private investors. Proceeds were used to pay for lands compulsorily purchased from landlords (Tang 1954; C. Chen 1961).

Roughly 12% of U.S. aid supported human resources, including sending more than 3,000 students and workers from Taiwan to the United States for advanced technical education. Of these, nearly 500 were part of two large joint ventures that supported many local students in vocational education and engineering education. In the first, the U.S. government enlisted Pennsylvania State University to assist the National Taiwan Normal University to build a department of teacher training for vocational schools, to help them educate technicians or skilled workers such as metal workers, carpenters, and automobile mechanics. In the second, the KMT government renamed the Tainan school National Cheng Kung University, in honor of the Ming leader who had brought an end to Dutch colonialism in 1661/1668, and the U.S. government enlisted Purdue University to rebuild the school's curriculum to emphasize education in the engineering sciences.

The KMT government happily permitted economic initiatives by local companies and the production of local engineering technicians and engineers. It remained confident that mainlanders would remain in control by virtue of their higher education and control of large companies. Locals would have to focus on climbing from the underside. This would change, however, with the rise of electronics.

References

Anonymous. 1945a. "Gong ye ji shu ren hui min ri kai chuang li zong hui 工業技術人會明日開創立總會 (The Founding Ceremony of the Institute of Industrial Technologists Will Happen Tomorrow)." *Min bao* 民報 *(People News)*, Nov. 10, 1.

Anonymous. 1945b. "Gong ye ji shu ren hui zuo ri kai chuang li zong hui 工業技術人會昨日開創立總會 (The Founding Ceremony of the Institute of Industrial Technologists Was Held Yesterday)." *Min bao* 民報 *(People News)*, Nov. 12, 2.

Anonymous. 1945c. "Shi yi ji shu bao guo, tai wan ke xue zhen xing hui cheng li 誓以技術報國台灣科學振興會成立 (Vow to Contribute to the Nation, Formosan Association Founded)." *Min bao* 民報 *(People News)*, Oct. 12.

Anonymous. 1950a. "Ben hui hui yuan ming ce 本會會員名冊 (FAAS Membership Roster)." *Tai wan ke xue* 臺灣科學 *(The Formosan Science)* 4: 35–41.

Anonymous. 1950b. "Zhong gong hui hui xun 中工會會訊 (CIE News)." *Tai wan gong cheng jie* 臺灣工程界 *(Taiwan Engineering)* 3(6): 1, 12.

Anonymous. 1976. "They Keep Moving On." *Taiwan Today*, Apr. 1.

Carleton, David. 2002. *Landmark Congressional Laws on Education.* Westport, CT: Greenwood Press.

Chang, Jui-Cheng 張瑞成, ed. 1990. *Guang fu tai wan zhi chou hua yu shou xiang jie shou 光復臺灣之籌劃與受降接收 (Plan on Recovering Taiwan, Receiving Japanese Surrender and Taking over Taiwan)*. Taipei: Party History Committee of the Central Committee of the Chinese Kuomintang (中國國民黨中央委員會黨史委員會).

Chang, Kuo-Hui, Gary Lee Downey, and Bono Po-Jen Shih. 2020. "Talents First!: Wu Ta-you and Science Policy Infrastructures in the Republic of China (1927–1970)." *The Korean Journal for the History of Science* 42(2): 449–473.

Chang, Kuo-Hui 張國暉. 2011. "Zhui xun zhu ti xing de gong cheng she ji: Wei quanzhuan xingzheng zhixia detaiwangao tietu mushe ji 追尋主體性的工程設計: 威權政治轉型下的台灣高鐵土木設計 (Pursuing Independent Engineering Design: State Transformation and Inclusive Civil Design in the Early Taiwan High Speed Rail Project)." *Tai wan she hui yan jiu ji kan 台灣社會研究季刊 (Taiwan: A Radical Quarterly in Social Studies)* 85: 157–200.

Chang, Kuo-Hui 張國暉. 2013. "Guo zu ke wang de ju ling: Tai wan ke ji guan liao zhi li de zhong guo mai luo 國族渴望的巨靈: 台灣科技官僚治理的中國脈絡 (The Leviathan That Nationalism Longed for: The Chinese Background of Taiwan's Technocracy)." *Guo jia fa zhan yan jiu 國家發展研究 (Journal of National Development Studies)* 12(2): 73–132.

Chang, Tsung-Han 張宗漢. 1980. *Guang fu qian tai wan zhi gong ye hua 光復前台灣之工業化 (Taiwan's Industrialization before Retrocession)*. Taipei: Linking Publishing (聯經).

Chen, Cheng 陳誠. 1961 [1995]. *Tai wan tu di gai ge ji yao 臺灣土地改革紀要 (Land Reform in China)*. Taipei: Chunghua Book (中華書局).

Chen, Tsui-lien 陳翠蓮. 2002. "Qu zhi min yu zai zhi min de dui kang: Yi yi jiu si liu nian 'tai ren nu hua' lun zhan wei jiao dian 去殖民與再殖民的對抗: 以一九四六年「臺人奴化」論戰為焦點 (Decolonisation versus Recolonisation: The Debate over the "Enslavement of Taiwanese" in 1946)." *Tai wan shi yan jiu 臺灣史研究 (Taiwan Historical Research)* 9(2): 145–201.

Chen, Yi-Shuan 陳翊軒. 2014. "Guo jia jue se yu gong cheng gu wen ye de qi yuan yu fa zhan: Yi tai wan mou da xing gong cheng gu wen ji gou de fa zhan li cheng wei li 國家角色與工程顧問業的起源與發展: 以台灣某大型工程顧問機構的發展歷程為例 (The Development of Engineering Consultant Industry in Taiwan and the Role of State: a Case Study on a Large Engineering Consultant Institution)." Master Thesis, Graduate Institute of National Development, National Taiwan University.

Chen, Yi 陳儀. 1945. "Wu ci guo fu ji nian zhou: Chen zhang guan yan ci 五次國父紀念週: 陳長官演詞 (Chief Executive Chen's Speech on the Fifth Memorial Week for the National Father, the Second Part)." *Min bao 民報 (People News)*, Nov. 28, 2.

Cheng, Li-Ling 鄭麗玲, and Li-Chu Yang 楊麗祝. 2009. *Tai bei gong ye sheng de hui yi 臺北工業生的回憶 (Former Students' Memory of Taipei Industrial School)*. Vol. 1. Taipei: National Taipei University of Technology.

Cheng, Tze Woody 鄭梓. 2002. "Chu tan: Zhan hou tai wan chu qi gao deng jiao yu de chong jian yu fa zhan 初探: 戰後台灣初期高等教育的重建與發展 (Reconstruction and Development of Higher Education in Early Postwar Taiwan)." In *Cheng gong de dao lu: Di yi jie cheng gong da xue xiao shi xue shu yan tao hui lun wen ji 成功的道路: 第一屆成功大學校史學術研討會論文集 (The Road to Success: Conference Proceeding of NCKU History)*, edited by Department of History at NCKU, 179–202. Tainan: National Cheng Kung University.

Cheng, Yu-Fong 程玉鳳. 1996. "Cong ji shu yi zhuan kan zi yuan wei yuan hui de ren cai pei xun: Yi san yi hui pai wei li 從技術移轉看資源委員會的人才培訓: 以三一會派為例 (On Talent Training of National Resources Commission from the Perspective of Technology Transfer: A Case Study of the 31 Club)." *Guo shi guan guan kan 國史館館刊 (Bulletin of Academia Historica)* 20: 139–175.

Cheng, Yu-Fong 程玉鳳. 2009. "Guang fu chu qi tai tang de xiao shou wen ti: Shi wu wan dun di tang de lai long qu mo (1945–1947) 光復初期臺糖的銷售問題: 十五萬噸敵糖的來龍去脈

(1945–1947) (The Sales of Tai-Sugar during 1945–1947: An Investigation into the Cause and Effect of 150,000 Metric Tons of Sugar Obtained from the Japanese Government)." *Guo shi guan guan kan* 國史館館刊 *(Bulletin of Academia Historica)* 21: 47–97.

Zheng, Youkui 郑友揆, Cheng Linsun 程麟荪, and Chuanhong Zhang 张传洪. 1991. *Jiu zhong guo de zi yuan wei yuan hui (1932–1949): Shi shi yu ping jia* 旧中国的资源委员会 *(1932–1949):* 史实与评价 *(National Resources Commission in Republican China, 1932–1949: Historical Facts and Evaluation)*. Shanghai: Shanghai Academy of Social Sciences Press.

Chiang, Kai-Shek, and Philip J. Jaffe. 1947. *China's Destiny & Chinese Economic Theory*. New York: Roy Publishers.

Chinese Institute of Engineers, ed. 1946. *San shi nian lai zhi zhong guo gong cheng* 三十年來之中國工程 *(China Engineering in Thirty Years)*. Nanking: Chinese Institute of Engineers.

Chu, Jiang-Huai 朱江淮. 2003. *Tai ji di yi wei dian qi gong cheng shi: Zhu jiang huai hui yi lu* 朱江淮回憶錄*:* 台籍第一位電氣工程師 *(Memoirs of Chu Jiang-Huai: The First Taiwanese Electrical Engineer)*. Taipei: Chu Jiang-Huai Foundation.

Chu, Wan-wen 瞿宛文. 2017. *Tai wan zhan hou jing ji fa zhan de yuan qi: Hou jin fa zhan de wei he yu ru he* 台灣戰後經濟發展的源起*:* 後進發展的為何與如何 *(The Origins of Taiwan's Postwar Economic Development: Why and How It Developed)*. Taipei: Linking Publishing (聯經).

College of Engineering, National Taiwan University. 2002. *Guo li tai wan da xue gong xue yuan yuan shi* 國立台灣大學工學院院史 *(History of College of Engineering at National Taiwan University)*. Taipei: College of Engineering, National Taiwan University.

Department of Civil Engineering, National Taiwan University. 1995. *Tai da tu mu wu shi nian* 台大土木五十年 *(Fifty Years of Department of Civil Engineering at National Taiwan University)*. Taipei: Department of Civil Engineering, National Taiwan University.

Education Yearbook Compilation Committee of the Ministry of Education 教育部教育年鑑編纂委員會. 1948. *Di Er Ci Zhong Guo Jiao Yu Nian Jian* 第二次中國教育年鑑 *(The Second China Education Yearbook)*. Shanghai: Shang Wu Yin Shu Guan (商務印書館).

Education Yearbook Compilation Committee of the Ministry of Education, 教育部教育年鑑編纂委員會. 1957. *Di San Ci Zhong Guo Jiao Yu Nian Jian* 第三次中國教育年鑑 *(The Third China Education Yearbook)*. Taipei: Zheng Zhong Shu Ju (正中書局).

Guan, Xiao-hong 關曉紅. 2008. "Shu tu neng fou tong gui: Li ting ke ju hou de kao shi yu xuan cai 殊途能否同歸: 立停科舉後的考試與選材 (Can All Roads Lead to Rome? Examinations and Candidate Selection after the End of the Imperial Civil Service Examination System)." *Jin dai shi yan jiu suo ji kan* 近代史研究所集刊 *(Bulletin of the Institute of Modern History Academia Sinica)* 59: 1–28.

Han, Cheung. 2016. "Taiwan in Time: Taiwan's 'Great Leap Forward'." *Taipei Times*, Oct. 9.

He, Te-Hsun 何德軒. 1947. "Hui wu 會務 (Conference Affairs)." *Tai wan gong cheng jie* 臺灣工程界 *(Taiwan Engineering)* 1(1): 18–19.

Ho, Ming-sho 何明修. 2016. *Zhi li po sui de tuan jie: Zhan hou tai wan lian you chang yu tang chang de lao gong* 支離破碎的團結*:* 戰後台灣煉油廠與糖廠的勞工 *(A Fragmented Unity: Workers in Taiwan's Oil Refinery and Sugar Factory in the Postwar Period)*. New Taipei City: Rive Gauche (左岸).

Hong, Sao-Yang 洪紹洋. 2016. "Tai wan gong kuang gong si zhi min ying hua: Yi fen chang chu shou wei zhu de tao lun 台灣工礦公司之民營化: 以分廠出售為主的討論 (The Privatization of Taiwan Industrial and Mining Company: Focusing on the Selling of Individual Factories)." *Tai wan she hui yan jiu ji kan* 台灣社會研究季刊 *(Taiwan: A Radical Quarterly in Social Studies)* 104: 103–148.

Hsueh, Jen-Yang 薛人仰. 1990. "Tai wan jiao yu zhi chong jian 臺灣教育之重建 (The Reconstruction of Education in Taiwan)." In *Kang zhan shi qi shou fu tai wan zhi zhong yao yan lun* 抗戰時期收復臺灣之重要言論 *(Selected Important Documents on Recovering Taiwan 1937–1945)*, edited by Jui-Cheng Chang 張瑞成, 299–304. Taipei: Party History Committee of the Central Committee of the Chinese Kuomintang (中國國民黨中央委員會黨史委員會).

Hu, Chun-Fa 胡均發. 1980. "Xin feng wu shi zhou nian you gan 欣逢五十週年有感 (Some Thoughts about Formosan Association's 60 Years)." *Tai wan ke xue* 臺灣科學 *(Formosan Science)* 34(3–4): 8–10.

Japanese Asset Management Committee of the Taiwan Province Takeover Commission 台灣省接收委員會日產處理委員會. 1947. *Tai wan sheng jie shou wei yuan hui ri chan chu li wei yuan hui jie shu zong bao gao* 台灣省接收委員會日產處理委員會結束總報告 *(Final Report of the Japanese Asset Management Committee of the Taiwan Province Takeover Commission)*. Taipei: Japanese Asset Management Committee of the Taiwan Province Takeover Commission.

Jesiek, Brent K., and Yi Shen. 2012. "Educating Chinese Engineers: The Case of Shanghai Jiao Tong University during 1896–1949." In *Engineering Development and Philosophy*, edited by Steen H. Christensen, Carl Mitcham, Bocong Li and Yanming An, 123–143. Dordrecht: Springer.

Jin, Guantao 金觀濤, and Qingfeng Liu 劉青峰. 2004. "Cong ge wu zhi zhi dao ke xue sheng chan li: Zhi shi ti xi yu wen hua guan xi de si xiang shi yan jiu 從"格物致知"到"科學"、"生產力": 知識體系和文化關係的思想史研究 From *Gewu zhizhi* to *Kexue* and *Shengchanli*: A Study of Knowledge Systems and Culture Relationships from the Perspective of Intellectual History." *Jin dai shi yan jiu suo ji kan* 近代史研究所集刊 *(Bulletin of the Institute of Modern History, Academia Sinica)* 46: 105–157.

Kao, Shu-Yuan 高淑媛. 2016. *Tai wan gong ye shi* 臺灣工業史 *(Taiwan Industrial History)*. Taipei: Wunan (五南).

Kirby, William C. 2010. "Engineers and the State in Modern China." In *Prospects for the Professions in China*, edited by William P. Alford, William Kirby and Kenneth Winston, 283–314. London: Routledge.

Kirby, William C. 2000. "Engineering China: Birth of the Developmental State, 1928–1937." In *Becoming Chinese: Passages to Modernity and Beyond*, edited by Yeh Wen-hsin, 137–160. Berkeley, CA: University of California Press.

Li, Yung-Chih 李永志, ed. 2011. *Tai wan ke xue zhen xing hui ba shi nian hui shi: Jian lun du cong ming li shi di wei* 臺灣科學振興會八十年會史: 兼論杜聰明歷史地位 *(The Eighty-year History of the FAAS: Also Discussing the Historical Status of Tu Thung-ming)*. Taipei: Formosan Association of Advanced Science.

Liao, Yen-Hao 廖彥豪. 2013. "Tai wan zhan hou kong jian zhi li wei ji di li shi gan yuan: Zhong tan nong di yu shi di gai ge (1945–1954) 臺灣戰後空間治理危機的歷史根源: 重探農地與市地改革 (1945–1954) (The Historical Roots of Spatial Governance Crisis in Postwar Taiwan: Revisiting Agricultural and Urban Land Reforms, 1945–1954)." Master Thesis, Graduate Institute of Building and Planning, National Taiwan University.

Lin, Ching-Po 林清波, ed. 2012. *Tai wan ying zao ye bai nian shi* 臺灣營造業百年史 *(Taiwan's Construction Industry: A Centenary History)*. Taipei: Yuan-Liou (遠流).

Lin, Chung-hsi. 2009. "The Silenced Technology—the Beauty and Sorrow of Reassembled Cars" *East Asian Science Technology and Society: An International Journal* 3(1): 91–131.

Lin, Hso-Mei 林秀美. 1999. "Tai bei di guo da xue zhi chuang she 台北帝國大學之創設 (The Establishment of Taipei Imperial University)." In *Tai bei di da de sheng huo* 臺北帝大的生活 *(Life at Taipei Imperial University)*, edited by Hsiu-Jung Chang 張秀蓉, 63–75. Taipei: National Taiwan University.

Lin, Ting-Sheng 林挺生. 1951a. "Lun tai wan zhi gong ye 論臺灣之工業 (On Taiwan's Industry)." *Tai wan gong cheng jie* 臺灣工程界 *(Taiwan Engineering)* 4(3): 70–71.

Lin, Ting-Sheng 林挺生. 1951b. "Tan tai wan gong ye 談臺灣工業 (On Taiwan's Industry)." *Tai wan ke xue* 臺灣科學 *(Formosan Science)* 5 (1–2): 8–9.

Lin, Yu-Sheng 林裕盛. 2014. “Guo jia yu qi ye gong sheng guan xi de xing su yu xiao shi: Yi zhong ding gong cheng gong si zhi fa zhan li cheng wei li 國家與企業共生關係的形塑與消逝: 以中鼎工程公司之發展歷程為例 (The Symbiotic Relation between the State and Enterprise in Taiwan: A Case Study on the Development of CTCI Corporation).” Master Thesis, Graduate Institute of National Development, National Taiwan University.

Ling, Hung-Hsun 淩鴻勛. 1961. “Wu shi nian lai de zhong guo gong cheng shi xue hui 五十年來的中國工程師學會 (The Chinese Institute of Engineers over the Past Fifty Years).” In *Jin ban shi ji yi lai gong cheng fa zhan gai kuang* 近半世紀以來工程發展概況 *(Overview of Engineering Development in the Past Half Century)*, edited by Dah-Her Huang 黃達河, 1–28. Taipei: Chinese Institute of Engineers.

Liu, Su-Fen 劉素芬. 2005. *Li guo ding: Wo de taiwan jing yan* 李國鼎: 我的台灣經驗 *(Kwoh-Ting Li: My Taiwan Experience)*. Taipei: Yuan-Liou (遠流).

Lu, Fang-sang 呂芳上, et al. 2000. *Du shi ji hua qian bei ren wu fang wen ji lu* 都市計畫前輩人物訪問紀錄 *(Records of Interviews with Pioneers in Taiwan's Urban Planning)*. Taipei: Academia Sinica.

Ou, Su-Ying 歐素瑛. 1997. “Guang fu chu qi tai wan zhi ye jiao yu zhi yan jiu (1945–1949) 光復初期臺灣職業教育之研究(1945–1949) (Professional Education of Taiwan (1945–1949)).” Master's Thesis, History, National Taiwan Normal University.

Statistics Office of the Chief Executive's Office of Taiwan Province 臺灣省行政長官公署統計室. *Tai wan sheng wu shi yi nian lai tong ji ti yao* 臺灣五十一年來統計提要 *(Statistics Summary of Taiwan Province over the Past 51 Years)*. Taipei: Statistics Office of the Chief Executive's Office of Taiwan Province.

Sun, Yat-sen 孫中山. 1920. *The International Development of China*. Shanghai: The Commercial Press.

Sun Yueqi 孙越崎. 1988. “Wo he zi yuan wei yuan hui 我和资源委员会 (The National Resources Commission and I).” In *Hui yi guo min dang zheng fu zi yuan wei yuan hui* 回忆国民党政府资源委员会 *(Recounting KMT Government's National Resources Commission)*, edited by History and Learning Committee Literature, CPPCC. Beijing: Chinese Literature and History Press.

Taiwan Historica, ed. 1995. *Kang zhan yu taiwan guang fu shi liao ji yao* 抗戰與臺灣光復史料輯要 *(Collection of Historical Materials of the Resistance War and Taiwan Retrocession)*. Nantou, Taiwan: Taiwan Historica.

Tang, Hui-sun. 1954. *Land Reform in Free China*. Taipei: Chinese-American Joint Commission on Rural Reconstruction.

Tu, Tsung-ming 杜聰明. 1948. “Kao cha nei di ke xue she shi yu yi xue jiao yu zhi guan gan 考察內地科學設施與醫學教育之觀感 (Some Thoughts after Visiting the Science Facilities and Medical Education in Inner Land).” *Tai wan ke xue* 臺灣科學 *(Formosan Science)* 2(2): 1–3.

Tu, Tsung-ming 杜聰明. 1950. “Tai wan sheng ke xue zhen xing hui zhi chong xin gai zu ji jin hou zhi shi ye ji hua 臺灣省科學振興會之重新改組及後之事業計畫 (FAAS Reorganization and Its Future Plan).” *Tai wan ke xue* 臺灣科學 *(Formosan Science)* 4(1): 1–2.

Wang, Jing 王璟, ed. 1984. *Zhong guo gong cheng shi xue hui qian tai fu hui san shi nian hui wu ji yao* 中國工程師學會遷台赴會三十年會務紀要 *(Minutes of the Thirty-year Meeting of the Chinese Institute of Engineers after Relocation to Taiwan)*. Taipei: Chinese Institute of Engineers.

Wu, Jao-keng 吳饒畊 (吳濁流). 1946. “Ti chang zhi ye jiao yu 提倡職業教育 (Promoting Vocational Education).” *Min bao* 民報 *(People News)*, Nov. 24, 1.

Wu, Zhaohong 吴兆洪. 1988. “Wo suo zhi dao de zi yuan wei yuan hui 我所知道的资源委员会 (What I Know about the National Resources Commission).” In *Hui yi guo min dang zheng fu zi yuan wei yuan hui* 回忆国民党政府资源委员会 *(Recounting KMT Government's National Resources Commission)*, edited by CPPCC History and Learning Committee Literature, 63–141. Beijing: Chinese Literature and History Press.

Wu, Tsong-Min 吳聰敏. 2023. *Tai wan jing ji si bai nian* 台灣經濟四百年 *(A 400-Year History of Taiwanese Economy)*. Taipei: Spring Hill (春山).

Wu, Yu-Ai 吳玉愛. 2016a. "Jiao da dian zi yan jiu suo di yi jie bi ye dian li 交大電子研究所第一屆畢業典禮 (The First Graduation Ceremony of the Institute of Electronics of Chiao Tung University)." National Chiao Tung University, https://news.lib.nycu.edu.tw/category/university-history/. Accessed Sep. 21, 2023.

Wu, Yu-Ai 吳玉愛. 2016b. "Jiao da dian zi yan jiu suo di yi jie yan jiu sheng ru xue ji 交大電子研究所第一屆研究生入學記 (Admission Record of the First Class of the Institute of Electronics, Chiao Tung University)." National Chiao Tung University, https://news.lib.nycu.edu.tw/category/university-history/. Accessed Sep. 21, 2023.

Yen, Chao Pang 顏朝邦. 1950a. "Bian ji za ji 編輯雜記 (FAAS News: From the Editor)." *Tai wan ke xue* 臺灣科學 *(The Formosan Science)* 4: 42.

Yen, Chao Pang 顏朝邦. 1950b. "Hua nan kao cha ji 華南考察記 (Report of Visit to Southern China)." *Tai wan ke xue* 臺灣科學 *(The Formosan Science)* 3(1): 1–6.

Yen, Yen-Chen 嚴演存. 1989. *Zao nian zhi tai wan* 早年之臺灣 *(Taiwan in the Early Years)*. Taipei: China Times Publishing (時報出版).

Yu, Ching Chun 余慶俊. 2009. "Tai wan cai jing ji shu guan liao de ren mai yu pai xi (1949–1988) 台灣財經技術官僚的人脈與派系(1949–1988年) (Connections and Factions of Taiwan's Economic and Financial Technocrats, 1949–1988)." Master Thesis, Institute of Taiwan History, National Chengchi University.

4 Animating Two Taiwans

In 1965, several alumni of Shanghai Jiao Tong University who had left China for the United States before or during the War of Resistance committed to help the Republic of China build an electronics industry on the island province. The mainlander engineers agreed to share learning about industrial development that they had gained in their respective career trajectories. An early contributor, Tsao Tseng-Chueh (趙曾玨; 1901–2001), traveled briefly to Taipei from New York the previous year to deliver the lecture "Modern Industry and Its Development" (T. Tsao 1985, 55). Following his undergraduate work, Tsao had completed a degree in electrical engineering at Harvard University, returned to China where he held a number of appointments, and then left for the United States again to work at Consolidated Edison in New York. At the time of his lecture, he was serving at Columbia University as senior research engineer in its Electronics Research Laboratory (T. Tsao 1964, 8–9).

Tsao had earlier also helped the KMT government begin restoring Chiao Tung University on the island province as National Chiao Tung University Institute of Electronics" (國立交通大學電子研究所). Chiao Tung had been known among engineers in China as its MIT, yet few of the university's resources had traveled across the Taiwan Strait after the war (Anonymous 1966). During the commemoration of its 60th anniversary in 1956, a telegram arrived from the Chiao Tung Alumni Association of New York. It proposed to establish "an institute of electronics that will undertake the development of the transportation and communications enterprises"(Anonymous 1956), both of which are encompassed by the phrase chiao-tung (交通) in modern Chinese. When the KMT's Ministry of Education endorsed the plan, the national university officially restarted in the province as an institute of electronics (Anonymous, 1958) (Fig. 4.1).

With leadership from Tsao, the former school's New York alumni association actively steered the development of curricular contents for the institute through its Education Policy Committee. It was clear from the beginning that technological teaching and industrial development through the institute aimed to protect and promote "Free China," i.e., the

K.-H. Chang et al., *Engineers and the Two Taiwans*, Synthesis Lectures on Global Engineering, https://doi.org/10.1007/978-3-031-59766-4_4

Fig. 4.1 Institute of Electronics, National Chiao Tung University, 1966 (Courtesy of National Yang Ming Chiao Tung University)

Republic of China in the island province. One goal was to produce graduates who would stay in the ROC, not continuing the "flowing one-way out abroad" as they themselves had done. To that end, the educational design process prioritized "feedback" (反饋) from industry to align with their directions and material interests. Tsao and others argued against preparing students to do open-ended research, the emphasis in the United States, and in favor of "concentrat[ing] all our available resources" in areas with the "highest potentiality of marketing in the U.S." for research outputs and, hopefully, actual products and services (T. Tsao 1964, 4–5).

The first step, explained Tsao in a letter to its director, should be "training … [e]ngineers and [t]echnicians in semiconductor technology" through a "kick-off program of transistor and diode manufacture" on-site. The institute should then spend the next five years "catching up with the modern trend of electron device development," with a "final goal of manufacture and export" (T. Tsao 1964, 6–7). The institute already had in place a high vacuum facility that could be used to get started. Funding could come from the United Nations. Tsao estimated that the international market for the manufactured products of integrated circuits could be as much as a billion USD by 1970. Other Chiao Tung graduates working in New York at Syracuse University and Westinghouse Corporation endorsed the plan, as one professor explained:

> I would like to voice my support of your selection of integrated circuits as a particularly appropriate project on which the Chiao Tung Institute of Electronics is to devote its efforts.

> Integrated circuits are a new art, and much interest has been generated in their research, development and manufacture because of their important potentials. It certainly makes sense for us to try to catch up head-on with the frontiers of electronic developments which have obvious practical applications than to start from scratch on some difficult project with little hope of even keeping up with the relevant technology or of putting it to use. The existence of high-vacuum facilities at the Institute also makes it more feasible to start the work on integrated circuits (T. Tsao 1964, 8).

In Chap. 1, we mentioned briefly that the KMT government had initially not prioritized the development of manufacturing and service industries while transferring the project of national reconstruction to Taiwan. It considered the new province a temporary base of operations. Instead, it allowed a private industrial sector to develop by inviting U.S., Japanese, and European companies to come build factories, promising them low wages, virtually no environmental regulations, and favorable trade policies (Lin and Chen 2021, 398–400; T.M. Wu 2023, 422–439). Companies such as General Instruments, Philco, Philips, RCA, Texas Instruments, Motorola, Sony, Sharp, and Sanyo all took advantage (Amsden and Chu 2003; Hong 2016; S. Liu 2012 [1992], 371; T.M. Wu 2023, 406–420). This changed in the 1950s and 1960s as the government began focusing new attention on industrial development in the island province, especially in the field of electronics. Its interest was prompted by advice and pressure from the U.S. government to develop a market economy and accelerated by an expected 1965 announcement that the U.S. government would cease foreign aid to the Republic of China (Chang et al. 2020, 465; Jacoby 1967, 138–139; S.F. Liu 2005, 128–129; T.M. Wu 1988, 156).

The episodes in this chapter and the following chapter call attention to the different pathways taken by mainlander and islander engineers into and through the electronics industry. Until this period, engineers in Taiwan primarily came from elite groups: foreign engineers during the late Qing dynasty, Japanese engineers during the colonial period, and mainland engineers during the initial decades of the post-war period. It would be through the unique, highly-distributed organization of the electronics industry that local engineers would come to play a significant role. The rise of local engineers and their technical contributions to Taiwan and its people would stand apart from the identities and commitments of mainland engineers. Through the material work of building novel infrastructures for an emerging industry, electronics engineers from the 1960s to the 1980s would play key roles in animating not just one Taiwan, but two.

Mainlanders lead through METS, ITRI & EIRC

Tsao Tseng-Chueh also recruited members of the New York chapter of the Chinese Institute of Engineers to advise and lobby the ROC leadership to pursue the development of new industries on the island province. Also mainlanders, members of the New York chapter had earlier demonstrated support of the KMT government by providing technical advice during the war with Japan. Now they agreed to help transport to the ROC knowledge in engineering and engineering management by establishing in Taipei what became known as the Modern Engineering and Technology Seminar (METS). Each lasting two weeks, the METS seminars addressed a range of topics requested by advocates in the province. The New York chapter found speakers, and hundreds of engineers and government officials on the island attended. Five seminars took place during the first eight years. While topics ranged from petrochemical engineering to mining, irrigation to aerospace engineering, engineering education to business administration, and construction engineering to urban planning, a key technical subject at every seminar was electronics engineering (Y. Shih 1985; J. Wang 1985) (Fig. 4.2).

The mainlander president of the New York chapter, Pan Wen-Yuan (潘文淵,1912–1995), was soon hooked. Following his training in electrical engineering at the Shanghai campus of Chiao Tung, Pan had completed a Ph. D. in electrical engineering at Stanford University in 1940, worked during the war at Harvard University, and then joined RCA afterwards as a research scientist. By the time the seminars began, he had risen to direct RCA's David Sarnoff Laboratories. Pan traveled to Taipei to attend the first seminar,

Fig. 4.2 Modern Science and Engineering Seminar, 1968 (Courtesy of Pan Wen-Yuan Foundation)

Fig. 4.3 Modern Engineering and Technology Seminar, 1968, engineers and other experts arriving from the United States (Left second—Lin Ting-Sheng; third—Tsao Tseng-Chueh; fourth—Pan Wen-Yuan) (Courtesy of The Central News Agency, Taiwan)

convened the second, and made himself a key New York contact for subsequent editions, as well as for other initiatives (Pan 1985) (Fig. 4.3).

While attending the first seminar, Pan delivered a commencement address to graduates from the Chiao Tung Institute of Electronics, making a sustained argument for designing and manufacturing electronics products. He implored them to focus on revenue for the ROC. Their education at the institute had provided them with a "good tool," and all graduates had clearly demonstrated the "dedication and determination" necessary to "turn out good products of high quality and good workmanship." But, he asserted, those products "must be sold to consumers." Pan explained that it was not enough for the ROC to achieve rates of economic growth equal to West Germany and Japan because the island-based country was starting so far behind. Along with "prosperity," calculated as percentage growth, graduates must also see "wealth," an "absolute value." It had been sixteen years since the ROC "moved its Capital to this island" whose limited wealth still came primarily from agriculture. Acknowledging implicitly that the ROC's future may not include recapturing the mainland, Pan asserted that "Taiwan's future expansion has to depend on industrial developments which has [sic] infinite possibilities and there is almost no limit in scope." Those developments most certainly depended upon a "close coupling between this country and the U.S.A. on modern engineering and modern technology," and promised to "lead to greater things of greater importance to a great nation," i.e., the Republic of China (Anonymous 1966).

Eight years later, Pan built on this argument at a morning meeting in a small Taipei breakfast shop where he proposed a dramatic new plan for the KMT government (K. Cheng 1993, 174–175). The plan was for the government to insert itself directly into the private sector to an extent that it had previously considered inappropriate. The means would be to pursue what was emerging internationally as the heart of the electronics industry – the mass manufacture of semiconductors with integrated circuits (IC) (T. Tsao 1985, 56–57). And the key first step would be to dramatically expand the efforts of the METS seminars to appropriate American knowledge and techniques by sending engineers to the United States for on-site, face-to-face learning. Funded by the government, young engineers would learn all the techniques of IC production and then return home to build an actual pilot plant. The ROC Minister of Economic Affairs was present at the breakfast and agreed to support it (Hu 2011, 13–14). That minister was the mainlander engineer Sun Yun-Suan (孫運璿; 1913–2006), later to become premier of the ROC.

When the KMT government first gained control of the island in 1945, it had sent Sun Yun-Suan to direct the electrical and mechanical department of its power company. Sun had been working on the mainland as an electrical engineer at a power plant owned by the government and managed by the huge government bureaucracy, the National Resources Commission. Born in a coastal province, Sun had been trained at Harbin Institute of Technology, a school in the northeast co-managed by China and the Soviet Union. He was primed for a supervisory position after the war as one of the Society of 31 mainlanders who had been dispatched to the United States in 1942 by the National Resources Commission to receive training in major sectors of American industry. He had just spent three years at the Tennessee Valley Authority obtained extensive engineering knowledge of hydroelectricity (Kirby 1990, 44–46; A. Yang 1989).[1]

Over the next 18 years, Sun worked his way up to general manager of Taiwan Power (Taipower), by then the top state-owned enterprise. He did leave the province for three years beginning in 1964, reportedly to seek more income and possibly to gain leadership credentials (A. Yang 1989, 95). He accepted an offer from the World Bank to go to Nigeria as CEO and General Manager for its government-owned Electricity Corporation. During that three-year period, he recruited engineers from the ROC to help expand Nigeria's power infrastructure. He and his engineers proved their mettle as the corporation increased its power supply by 87% (A. Yang 1989, 105). Sun returned to the ROC in 1967 to become Minister of Transportation and Communications and, two years after that, Minister of Economic Affairs.

[1] The Society of 31 benefitted from U.S. President Roosevelt's Lend-Lease program during the war. Sun was one of seven who moved to Taiwan afterwards. Six achieved leaderships positions in the KMT government, including three leaders of state-run industries, two Cabinet officers, and one finance official.

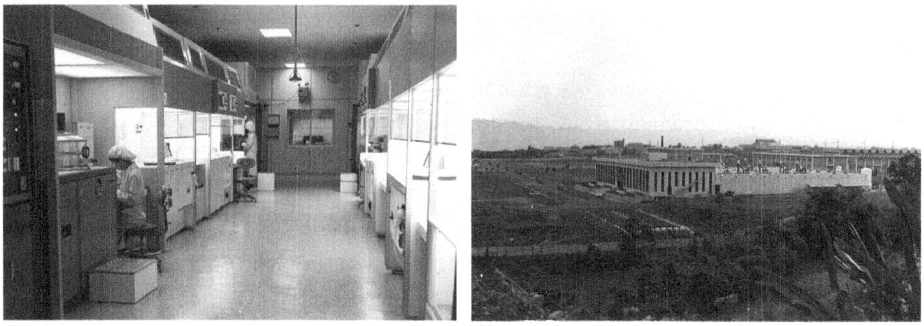

Fig. 4.4 Electronics Industry Research Center (EIRC), Industrial Technology Research Institute (Courtesy of Pan Wen-Yuan Foundation and ITRI)

Sun's return and positioning in government both illustrated and enacted its emerging commitment to develop technology-intensive industries on the island. In 1973, Sun led a process to combine three government institutes into a new research infrastructure, the Industrial Technology Research Institute (ITRI), with a mandate to ground new companies through research and development (K. Cheng 1993, 127–128) (Industrial Technology Research Institute 2013, pp. 12–13). And a year later it was he who attracted the mainlander Hu Ding-Hua to serve as deputy director of a new research infrastructure targeting electronics, the Electronics Industry Research Center (EIRC) (Fig. 4.4).

Pan Wen-Yuan helped to coordinate from the United States, leading the creation of the Technical Advisory Committee to support EIRC (Pan 1974). The advisory committee recommended that EIRC seek technical support from seven American companies: Hughes Aircraft, RCA, AMI (American Microsystems, Inc.), Sprague, General Instruments, Intersil, and Fairchild (J. Chang & Pan Wen Yuan Foundation 2006, 89–90; S. Wu 2019, p. 87). To implement the plan to manufacture semiconductors, Pan helped EIRC finalize a $350 million USD contract with RCA, his employer, to train three dozen engineers in design, manufacturing, packaging, testing, and production management and then support the construction of a pilot plant. A first batch of 19 was sent in April, 1976, followed by the remaining 17 later that year (Industrial Technology Research Institute 2013, 18–19). In order to participate in the project without any conflicts of interest, Pan retired early from RCA, giving up many retirement benefits (Pan Wen Yuan Foundation 2006, 4).

In the process of fulfilling his plan, Pan persuaded his wife to resign from a tenure-track professorship. Remembering the difficult decision in a later interview, she tied it to Pan's sense of duty. "Over many years, Wen-Yuan and I often discussed 'how to help his great mother country,'" she reported, and "before he became too old to do anything this was his biggest wish." In early 1975, "we convened a family meeting to discuss" his intent "to give up his managerial position as soon as possible." And "because he needed my company to fly to Taiwan often," he "wanted me to give up [the] professorship … I [had held] for 18 years." Although early retirements would bring a "big financial loss," "finally I agreed with him to put national devotion prior to personal happiness…" (J. Chang & Pan Wen Yuan Foundation, 28–29).

The New York chapter's advisory committee helped EIRC locate and select participants and then choose which RCA divisions they should visit. Hu Ding-Hua served as project leader. Four teams in the first batch went off to RCA plants, distinguished by the steps in the manufacture of semiconductors (Pan Wen Yuan Foundation 2006, 4). One group with four members went to New Jersey to learn IC design. A group of seven went to Ohio to learn IC manufacturing. A team of three went to California for IC testing and packaging. And a team of four went to Florida to learn about IC equipment.

The so-called "RCA 19" included roughly half mainlanders and half islanders. Three of them (two mainlanders and one islander) were pursuing Ph.D. degrees in the U.S. at Princeton University. One mainlander reported that he learned about the project from the *Central Daily* newspaper, which he and all other KMT members studying in the United States got for free (S. Wu 2019, 83; D. Yang 2011, 11). Most of them had been working in the province. Their average age was 28 years old (Chang et al. 2006, 45; S. Wu 2019, 86) (Fig. 4.5).

After returning from the U.S., not all the RCA 19 went to work building the pilot plant. All those who did, reported Hu Ding-Hua, were from mainlander families "whose parents had served in the military, education or government." As such, he asserted, they maintained "a family belief that national development was important" (Chang et al. 2006,

Fig. 4.5 Some RCA 19 members and an RCA manager (Courtesy of ITRI)

133–134). Some members of the RCA 19, including a number of the islanders, left the government organization to support small family businesses. Hu speculated that "they might have believed that the plant would not be successful." As we will see below, their time spent in the U.S. could also have made more visible a separation that made national development for the ROC less than a first priority.

The EIRC-RCA contract required RCA to provide technical support until the pilot plant could reach a 17% yield rate, referring to the percentage of semiconductors on wafers that worked properly. Removing impurities is difficult. Yet within six months of its start-up in 1977, the plant achieved an astounding rate of more than 70%. It was also producing around 4,000 wafers per week, more than the comparable RCA plants in the U.S (S. Wu 2019, 111–112, 117–118, 120–121). Greatly impressed with the speed of this start-up, RCA actually sought to purchase the plant, but EIRC declined (S. Wu 2019, 122). Indeed, one RCA 19 engineer later asserted that all those remaining would have quit if the government sold the plant to RCA (Chang et al. 2006, 111; S. Wu 2019, 122–123).

Three years later, EIRC spun off its first large company, the United Microelectronics Corporation (United Microelectronics). The government-supported company built its first integrated circuit for a government agency, the military, with a patriotic purpose in mind: a timing device to explode balloons carrying propaganda flyers over the mainland (D. Yang 2011, 17). The company proudly named its first commercial integrated circuit CIC-001. It later renamed it CIC-0001 to signal an expectation of 9,998 more to come (S. Wu 2019, 99) (Fig. 4.6).

A government organization, EIRC assigned 65 of its engineers to train employees of United Microelectronics, and 38 employees went to work for new company to manage

Fig. 4.6 ITRI's first commercial IC used in an electronic watch, 1978 (Courtesy of ITRI)

and lead them (S. Wu 2019, 132, 136). One of those who transferred was the mainlander Robert Tsao (曹興誠, 1947–), one of the RCA 19. Born in China with a father in military service, Tsao had earned a B.S. in electrical engineering at Taida and a master of management at Chiao Tung, and then stayed on the island to work at ITRI. He served as lead design engineer on the first integrated circuit project.

United Microelectronics then went to work integrating the design and manufacture of specialized integrated circuits for service in a range of electronics devices and products (Industrial Technology Research Institute, 2013) with the intention of supporting a unique industrial infrastructure. The government's plan contrasted in an important way with the infrastructures of state-supported electronics companies that had already been growing in both Japan and Korea. The *zaibatsu* in Japan and *chaebols* in Korea aimed for vertical integration, i.e., the centralized coordination and control of subsidiaries by a parent corporation. UMC would instead manufacture a variety of integrated circuits that other companies could build upon in a range of different products. The others would include small, family-owned companies. That is, serving as a "seed to boost [a] whole IC industry," United Microelectronics would join with small companies to "guard our high-tech industrial development" ((J. Chang & Pan Wen Yuan Foundation 2006, 133–134; S. Wu 2019, p. 139). During its first decade, United Microelectronics supplied integrated circuits for such products as electric watches, telephones, video games, and personal computers made in a variety of settings (R. Tsao 1999, 122–123; 2004, 60).

This KMT vision and plan for distributed production affirmed and extended longstanding practices of hierarchical economic governance on the island, in this case accepting the presence of large numbers of small private companies organized by islander families around large, state-organized enterprises. During the late Qing dynasty merchants had found ways of elevating themselves through contractual relations with the provincial government. During the colonial period the Japanese government had planned for workers to fill low-level jobs in the government and private companies, but also permitted islander families to freelance around them through contractual relationships, especially to support the construction, sugar, and chemical industries (Kao, 2016; Lin 2012, 157–175). And during the early years of KMT rule, the mainlander government had encouraged islander family companies to multiply. These provided needed goods and services, including by subcontracting on large government-supported projects. And as small, distributed private businesses, the family companies did not raise political concerns nor contest the authority of the central government (Shieh, 1992, 41–43, 61–62, 124, 138–141, 216).

Extending these practices to electronics meant that the KMT wholly expected to maintain hierarchical power relations that would privilege state-supported enterprises and, hence, mainlander engineers. Islander engineers working in wholly local companies would be invited to help realize the new industry, supporting them from the margins while also remaining dependent on the state-owned organizations. Hu Ding-Hua, then still at EIRC, called the new model for the electronics industry in Taiwan "vertical disintegration" (C. Chang and Yu, 2001, 3–23).

Not surprisingly, UMC populated itself mainly with mainlander engineers, but with at least one important exception. Note that relatively few mainlanders had had any experience building new companies to operate in the private sector, and entering that messy arena posed a big risk. To serve as UMC's first managing director, EIRC found an engineer who had already experienced success in the IC industry in the United States and hired him even though he was an islander.

Tu Chun-Yuan (杜俊元,1938–2023), like Chew Tsai-Hsin above, had managed to gain admission to Taida in electrical engineering. Akin to so many mainlanders, he was able to get to the U.S. to pursue a Ph.D., in his case at Stanford University, and then stayed in the U.S. to work at the IBM Watson Research Center in New York. Tu had returned to the ROC in 1968 because of family issues and he gained positions teaching at both Chiao Tung and Taida In addition to his achieved status from working in the U.S., Tu's main local credential elevating him to visibility was success in building a small company to package and test integrated circuits in 1971. Among other things, packaging an IC in an encasement shields the electronic circuit from corrosion and damage and provides electrical contacts with external circuits, and testing calculates the reliability and yield of a batch of ICs. With packaging and testing being the finishing touches of IC production, Tu's company did not challenge the industrial hierarchy.

Mainlanders Lead Through Hsinchu Park And The Ten Major Construction Projects

The main government figure supporting the Modern Engineering and Technology Seminar was the senior official Li Kwoh-Ting (李國鼎, 1910–2001). Li had attended National Central University in Nanjing before studying physics at Cambridge University in the 1930s and then returning to the mainland to work in the nationalist government. When the government moved to Taiwan, he became president of the state-owned Taiwan Shipbuilding Company. During the 1950s, he served in the Economic Development Commission working on industrial development plans and then gained significant responsibility for steering U.S. aid as head of a development and investment center within the Council for U.S. Aid. When the U.S. announced in 1964 its intention to terminate aid to the ROC in 1965, he became a vocal activist for developing new industries. He was promoted to economic minister with that mission in mind, followed later by Sun Yun-Suan.

While the government's initiative in electronics relied heavily on engineers with ties to Chiao Tung University in Shanghai, Li and Sun also undertook a second initiative that drew heavily on engineers with ties to National Tsing Hua University in Beijing. When Chiao Tung University was restarted in 1956 as National Chiao Tung University, National Tsing Hua University was restarted and located also in Hsinchu. Its fourth president, Hsu Hsien-Hsiu (徐賢修, 1912–2001), was a mathematician who strongly supported education in engineering because of its potential economic benefits. After completing degrees in mathematics at Tsing Hua and Brown University, Hsu had served on the faculties of the Illinois Institute of Technology and Purdue University in the United States, and had served actively in the Chicago-based alumni association for Tsing Hua. After accepting Chiang Kai-Shek's invitation to become the school's president, he formalized the addition of a college of engineering with a focus on two projects: electric vehicles and geothermal power.

In 1973 Hsu transferred to a senior position in the National Science Council with a mission to build science-based industrial parks. The idea was to attract foreign investors to establish and upgrade a range of industries across the island. A 1975 visit to Toyota and Seiko in Japan initiated a plan to build one in Hsinchu. The plan garnered strong support from then-Premier Chiang Ching-Kuo, son of Chiang Kai-Shek, who encouraged Sun Yun-Suan to move ahead (H. Lin 2021, 406–407). Hsu invited other professors from Tsing Hua University to join the project, including the chemist Liu Chao-Shiuan (劉兆玄, 1943–) who had established a new department of planning in the National Science Council (H. Wang 2012, 146–147, 157–159). Hsu also traveled to the United States to promote the planned park, asserting that Taiwan was training lots of engineers and scientists who could staff new divisions on the island and would accept low salaries (Fig. 4.7).

Fig. 4.7 Sun Yun-Suan (right) visited the first IC foundry, 1977 (left—Hu Ding-Hua) (Courtesy of The Central News Agency, Taiwan)

The science park plan for Taiwan also differed in a crucial way from a similar plan then underway in the Republic of Korea during the regime of Park Chung-hee. Park's plan contracted directly with 29 elite families to establish industrial conglomerates, the *chaebols*, funded by the sale of lands and direct support from the Economic Planning Board (Han and Downey 2014, 79–87). Such a plan could not fit Taiwan, however, because of the separation between mainlander and islander families. Given their privilege and access to governance, mainlanders were generally less interested in private business than islanders, and their numbers were far fewer. At the same time, the government could not easily bypass mainlander families in favor of support for islander families—yet. So rather than assigning projects to families via a centralized process akin to Korea, the Hsinchu science park administration requested formal proposals, offering access to space, financial support, and a range of incentives. And early on, approvals for proposed projects often went to entrepreneurial mainlanders working overseas (C. Chang 2011, 7; P. Ho, 2004, 191, 196–197, 221).

With an initial emphasis on precision machinery and biotechnology, the park in Hsinchu struggled at first to attract viable projects and tenants. It then grew quickly during the 1980s through a turn to electronics. By 1990, it had become one of the most advanced centers for industrial research and development in Asia. And of the 141 companies there in 1992, 80% were in integrated circuits and information and computer technologies (ICT). Indeed, between 1983 and 2002, 90% of the revenues from the park came from companies focused on computers and computer peripherals (D. Yang and Chen, 1996).

Throughout the early period, mainlander engineers with connections to National Tsing Hua University figured highly in the science park leadership, along with alumni from the National University of Amoy (later known as Xiamen University) on the mainland who had gone abroad before 1949. A key early figure was Irvin T. Ho (何宜慈, 1921–2003), who suggested the focus on electronics and was invited by Hsu Hsien-Hsiu to lead the National Council's initiative to realize it. Ho had completed a bachelor's degree in electrical engineering at Amoy University and then, with the support of the Amoy president, Sah Pen-Tung (薩本棟), went on to earn his Ph.D. at Stanford University. Sah was himself a Tsing Hua graduate who, in the 1920s, also completed a master's degree at Stanford University and then a Ph.D. at Worcester Polytechnic Institute in the U.S. before returning to Tsing Hua as professor and then moving on to Xiamen (or Amoy in the local dialects). For his part, Ho remained in the U.S. after graduation, working at Fairchild Semiconductor International Inc. and then IBM, where he stayed for 16 years before joining the National Science Council's project on training and education in electronics.

The Chiao Tung and Tsing Hua initiatives also gained support from another large-scale initiative designed to ground industrial development—the so-called "Ten Major Construction Projects." Prior to its termination, U.S. aid had grounded infrastructural projects to support the expanded population living on the island. These included, in particular, electric power systems and telegraphy. With the termination of U.S. aid and vulnerabilities exposed by the 1973 oil crisis, Premier Chiang proposed the ten projects to ground the government's emerging commitment to industry on the island itself, to be completed by 1979. The ten included a nuclear power plant, an integrated steel mill, an oil refinery, a shipyard, an international airport, a highway project, two railway projects, and two ports (Figs. 4.8 and 4.9).

Chiang Ching-Kuo became President of the Republic of China the same year that the United States stopped recognizing the ROC as the government of China. Put overly succinctly, Chiang authorized a turn to the island as more than just a temporary host for the nationalist government (H. Lin 2021, 391–392). Any plan or dream to retake the mainland basically disappeared with the loss of U.S. recognition. The KMT government had to define a relationship with Taiwan that was more than a province of China.

In addition to greater openness and tolerance of dissent, relaxed controls on media and speech, and increased participation in government by Han and Hakka islanders, Chiang's turn imagined Taiwan as a center for industrial development. Chiang actually had some understanding of industrial projects from his early education and work experiences in the Soviet Union (Taylor 2000, 56, 61–63, 66). Although there were concerns about inflation from extensive spending, growth in the island's economy during the 1970s did affirm the value of the investment. And the Republic of China became firmly attached to the island of Taiwan (H. Lin 2021, pp. 391–392).

Fig. 4.8 A ten construction project: National Highway, 1977 (Courtesy of Taiwan Cultural Memory Bank)

Fig. 4.9 A ten construction project: North Link Railway, 1980 (Courtesy of Taiwan Cultural Memory Bank)

Islanders Fill the Gaps

Sun Yun-Suan's leadership as Minister of Economic Affairs (1969–1978) and then Premier (1978–1984) mattered. He actively encouraged the development of both large state-owned or state-supported companies, run primarily by mainlanders, and smaller, wholly private companies operating independently of the state—an arena heavily populated with islanders (Y. Wu 2005).[2] This disposition was not new. When Sun first came to Taipower in 1947, he had to figure out how to quickly replace 3,000 Japanese engineers and technicians who were forced to leave. His plan included recruiting 300–400 islander engineers-in-training and engineering-technicians-in-training from the technical schools in Taipei and Tainan. As he later put it while serving as minister, "All the entire nation's people who can do things are my people" (全國能做事的人都是我的人) (A. Yang 1989, p. 87).

Sun's inclusive approach to industrial development turned out to fit the material contents of the emerging electronics industry. Electronics technologies quickly became

[2] Chiu Chun-Pang, interview with Chang Kuo-Hui, April 13, 2018.

known across Taiwan and elsewhere as highly-distributed, heterogeneous networks, leaving ever-expanding and multiplying openings to be filled.[3] At levels that far exceeded the factory-based infrastructures in established industries, these gaps offered many opportunities for participation in design and manufacturing. Pursuing those career prospects generated a rapidly increasing array of devices and connections that linked together diverse electrical and mechanical functions.

The semiconductors on a computer circuit board, for example, act both as many distinct devices and as just one device among many that are necessary for humans to interact with them and use them to process digital information. And at the far end of what would become a densely-populated continuum, electronic communications systems span large expanses of territory and include enormous complexes of interacting techniques, devices, infrastructures, and people. At every scale, a given electronics network could become a standalone device or just one component of yet another engineered network.

Only large companies possessed the resources necessary to build and maintain the fabrication plants for semiconductors, aka the foundries. Committing a company to fabrication meant also committing to massive research and development: increasing wafer sizes, decreasing the sizes of transistors etched onto wafers, increasing production capacity, and achieving ever-higher yield rates. At the same time, the vast majority of electronics companies in Taiwan actually started off as small enterprises of one sort or another, usually funded by short-term investments from relatives and run by families rather than professional managers (Chiu 2015, 96–98; Johnstone 1994; Lam 2011; S. Shih 1996, 40–41). The family-owned companies could take advantage of the diversity and heterogeneity of IC-driven devices to help semiconductors travel quickly into virtually all existing industries.

At first, those families were mostly mainlanders taking advantage of openings in the private sector that opened up with Chiang Ching-Kuo's rise to leadership positions (Prime Minister 1971–1978; President 1978–1988). Unitron, the small company that helped Chew Tsai-Hsin to redirect his career away from foreign employment, was an example. Its founder, Simon Sze (施敏, 1936–2023), was a mainlander.

Although he grew up in a low-income family, Sze had made his way through Taida in electrical engineering and then to the United States for his Ph. D. at Stanford University, followed by a research career at Bell Labs. In 1966, the U.S. government's imminent separation from Taiwan motivated his short-term visit to teach at National Chiao Tung University, and then the next year he joined a government task force to support development of the new industry. He had learned in the process that graduates in electrical engineering who stayed in the country often had to work for foreign-owned companies because there were no local companies to hire them. "I wanted to help Taiwan

[3] In this account, we refer to electronics broadly to include semiconductors, electrical appliances, software and hardware, telecommunications, and data processing equipment (Amsden & Chu, 2003, pp. 174–175).

build semiconductor industry [sic]," Sze later wrote. "It was the right industry for Taiwan since Taiwan lacked natural resources and capital but had lots of talents. (L. Wang 2013, p. 225)" Sze was able to find funding of one million USD from a friend of his father-in-law who did business in the textile industry (Department of Electronics Engineering 2008). The new company was too small to become a central node, producing integrated circuits itself, so like Tu Chun-Yuan, Sze positioned the business on packaging and testing semiconductors (L. Wang 2013, 242).

Chew Tsai-Hsin was an attractive addition to the company because, although an islander, he had enhanced his knowledge and status through international employment. After joining Unitron, Chew founded a department for research and development on semiconductors that would extend the company's work beyond packaging and testing to include design and product manufacturing. It was moving from filling gaps to establishing niches. Work in that department, for example, later grounded a project to manufacture and successfully commercialize the first electronic calculator in Taiwan (S. Shih 2018, 4–5).

In the late 1960s and early 1970s, an increasing number of islander families sought ways to enter and seek to climb through this emergent industrial infrastructure, just as many had tried to do under the Japanese and during the early days of KMT rule. This time, however, islanders were seeking entry to a novel industrial infrastructure that was growing with a distribution of expanding openings and new niches. The functioning of nodes depended upon the filling of openings, and relationships among them were changing fast.

As in the past regimes, the industrial initiatives of islanders typically started by providing supplemental support or production work at the margins. But as the networked industry expanded and grew, purported distinctions between centers and peripheries of production and associated hierarchies of mainlanders over islanders became increasingly blurred. Islander companies found ways of contributing directly to the manufacture of devices and products across a wide range of niches, elevating themselves in the process. Consider two examples.

Shih Zhèn-Róng (施振榮 1944–) was another islander who started his career at Unitron, having turned down a 1971 offer to work at Philips (S. Shih 2018, 4–5). He wanted to focus on front-end research and development (R&D) rather than support activities, making clear that his goal was to create new advanced technology for Taiwan. He worked with Chew to design, develop, and market an electronic calculator and then, a year later, helped establish a new company to produce a pen that could tell time. Four years after that, now as Stan Shih, he co-founded a small firm, Multitech, that distributed electronic parts and consulted about the design of integrated circuits before it turned to build its own microprocessor, the key control device in a computer. Along with an additional turn to product design and manufacturing, business expanded greatly and Multitech grew to become five companies under the name Acer. As Acer then expanded to become a top-five manufacturer and vendor of PCs and notebook computers worldwide, it also signaled the possibility that a company founded by islanders could shed the image of a small or marginal contributor to the electronics industry (Otellini 2006; S. Shih 2018) (Fig. 4.10).

Or consider Santron, founded by an electrical technician's son who had gained admission to Taida in electrical engineering as well as five other Taida graduates who also came from low-income families (two other islanders, two from Hong Kong, and one from Burma). In 1972, the group secured funding from a wealthy islander family that had hired the electrical technician to repair his home audio system. Focused initially on the design and manufacture of calculators, the six developed managerial practices for designing new products, managing production lines for consumer products, marketing products to buyers inside and outside of Taiwan, and coordinating the whole financially. But the first project did not work out as the patron wanted to focus on audio technologies and forced the six to leave.

With funding from another wealthy islander family, they started Kinpo Electronics, Inc., believing that initial work on calculators could ground forays into the burgeoning world of computing. "We also worked with Qume and AT&T (in the U.S.)," reported one founder. "At the time, monitors were using CRT technology and had pretty good gross profit. We transitioned from calculators to computers. It was a natural trend; they are all digital applications" (Lam 2011, 8). Four of them later went on to start companies that

Fig. 4.10 Acer (Multitech) factory in Hsinchu Science Park, 1982 (Courtesy of National Archives Administration)

became prominent makers of desktop and laptop computers and a variety of peripherals: Quanta Computer, Inventec Corporation, and GVC. Two stayed and helped it spin off Compal Electronics, Inc.

While cases such as these broke through the glass ceiling that had kept small companies small, the main story of islander family participation in electronics was simply, or not so simply, their multiplication. Writing in the 1990s about the rapid expansion of the industry in Taiwan, one prominent observer, Annalee Saxenian, called particular attention to small companies. She pointed out that the "IT industry" then consisted of 10,000 "highly specialized" producers, mainly small and medium-sized companies, located within a fifty-mile region between Taipei and the Hsinchu Science Park in northwest Taiwan (Saxenian 2006, 126–127). The vast majority of these companies were founded by islanders, who constituted a significant majority of the population.

In contrast with the large firms sponsored and supported by the government, the smaller ones spread out to fill gaps and establish niches. However, unlike family firms during the colonial and early KMT periods, they were not directly dependent on supervisors or contract agents that preserved hierarchical power relations. The heterogeneous distribution of the industry was unprecedented, and those that could grow were not prevented from growing. Rough-and-ready entrepreneurial positioning enabled small islander companies to become integral contributors to a rapidly globalizing industry with multi-national flows. Rather than serving large, predominantly mainlander, companies through subordinate positions, they were working in parallel and in conjunction.

Multiplying Engineering Technicians and Engineers

Both the large and small companies were hungry for technical talents [技術人才] to do the actual work. The mathematically- scientifically-intensive contents of much of the work limited the extent to which on-the-job training could suffice. Advocates for the new industry from all directions saw expanding the availability of technical education for engineering technicians and engineers as necessary to build it.

Both mainlanders and islanders had long-standing commitments to primary and secondary education as essential to human development, demonstrated at the secondary level through successful performance on exams (McCusker and Robinson 1962, 7). As early as the 1950–1951 academic year, there were 1,444 schools at those levels, and the number continued to grow quickly (Ministry of Education 1971, pp. 2–4). Still, it remained the case that dirty hands carried a stigma ("black hands") (McCusker and Robinson 1962, 22). The introduction of formal education for a shiny, seemingly clean new industry provided a happy convergence of a base commitment to learning, a continuing desire to avoid dirty work, and an opportunity to elevate islander families through gainful employment that promised ever-increasing income.

The Minister of Education provided a stimulus for initiatives in engineering education by commissioning the Stanford Research Institute (SRI) in the U.S. to "[e]xplore means to improve education and training programs to enable the country better to achieve its economic development plans." The resulting two-volume report in 1962 described and assessed existing educational practices through the lens of internationally-competitive economic development. Implicitly brushing aside practices of Chinese learning, it critiqued an over-emphasis on "mere intellectual training" and preparation for "less important vocations" that were "of minor importance to the country's socio-economic development." It lamented that the "most able young students tend to view technical and vocational schools as undesirable second or third choices" (McCusker and Robinson, 1962, 1, 14, 23, 47, 57).

The report argued for strengthening vocational secondary schools and technical institutes at the tertiary level. It proposed measures to "improve the prestige and attractiveness of economically important occupations" and "bring … education into closer support of socioeconomic development" (McCusker and Robinson 1962, 57, 20).[4] It argued, in particular, for increasing the number of institutes focused on educating engineering technicians. At the time, there were only the public Taipei Institute of Technology and the private Tatung Institute of Engineering. The engineering-related offerings at both schools had long emphasized civil engineering to support the large state-owned companies dedicated to basic physical infrastructures.

The external recommendations converged with emergent priorities in domestic policy (S. Chen 1991; Zheng et al. 1991). The KMT government almost immediately authorized the construction of both public and private junior colleges at the tertiary level to produce a range of trained technicians. It was 1964, the year that the United States government first signaled that it would be terminating direct aid to the ROC. Of the sixty-two junior colleges that were built during the next decade, twelve were public, 50 were private, and fully 30 of them aimed specifically to produce engineering technicians (S. Chen 1991).

KMT officials were well-aware that the primary beneficiaries of the new junior colleges would be islanders. Indeed, that was the point. The established hierarchy between state-sponsored companies controlled by mainlanders and small private companies operated by islanders made expanding technical education for islanders feel like an effort that would affirm the existing order, not threaten it.

The families of young islanders welcomed this initiative because, for the most part, their sons (and sometimes daughters) were unable to pass the entrance exams to get into a university. At least it was an opportunity for education beyond the secondary level. And for all those islander and mainlander elites who funded the private junior colleges, the initiative provided an opportunity to earn returns from a private initiative that carried official sanction. Still, the mainlander government provided relatively little assistance and oversight in curricular development and educational contents, and the quality of both the public and the private institutes remained poor for a significant period of time (S. Chen 1991; Huang, 2017).

[4] See also a UN report making similar recommendations, not discussed here (UNESCO, 1964).

By 1970, more than 106,000 students were enrolled in junior colleges of all types, including those aimed at supporting government administration and other official activities (Ministry of Education 1971, pp. 86–87). Yet remarkably, of that total nearly one third, or 34,000 students, were making their way through curricula categorized as engineering (Ministry of Education 1971, pp. 88–89). The vast majority were young, male islanders making their way into a new industry whose opportunities lay not only in large, state-owned companies.

During this early period, technical universities grew more slowly. The KMT government had previously had little incentive to build them. During the 1950s, it had established university programs for public administration, military, police, and Chinese learning, all to support the government's temporary stay on the island (S. Chen 1991; Huang 2017). And given the growing tension with the People's Republic on the mainland, there was reason for concern that elite university graduates would go abroad, especially those in engineering and the sciences (Greene 2008; T.y. Wu 1989).

By 1970, only four institutions offered bachelor's degrees in electrical and electronics engineering. The students at Chiao Tung and Tsing Hua were mostly mainlanders. The student body at Taida was in the process of shifting from being majority-mainlander to majority-islander. And National Cheng Kung University, a majority-islander school in the south, was turning to electrical and electronics engineering after long supplying engineers for the chemical industry in Kaohsiung and construction projects for the government. The government did add the National Taiwan Institute of Technology (later Taiwan Tech) in 1974 to enable select graduates of junior colleges and vocational secondary schools to seek higher degrees. Largely for the same reason, the government also promoted the private Tatung technical institute to university status. By 1980, the number of university graduates in engineering had increased to more than 7,200 from fewer than 300 in 1950. And fully one-third of all degrees in higher education went to engineers, increasingly in electrical and electronics engineering (Department of Statistics 1990).

Still, islanders graduating with bachelor's degrees in electrical and electronics engineering could not yet easily find jobs with domestic companies through the 1970s. Small companies mostly wanted technicians at first. Large companies did begin to provide opportunities for qualified islander engineers, but they mostly favored mainlanders in hiring and promotion (M. Ho 2016).[5] Islander graduates plotted a range of pathways, including employment in foreign companies, teaching in universities or junior colleges, and founding new companies themselves (C. Chang 2011; Shih 2011).[6]

[5] Ku Po-Kuang interviewed by Chang Kuo-Hui, January 5, 2020.

[6] Ku Po-Kuang interviewed by Chang Kuo-Hui, January 5, 2020.

Electronics Elevates Islander Engineers

Through the 1960s and 1970s, the electronics industry emerged in Taiwan as an infrastructure of large and small companies that needed one another to prevent subservience to outsiders. The government mobilized its China Productivity Center, established in 1955, to help the small family-run companies acquire practices of business administration and management (Hou and Gee 1993). The main means of connection and collaboration, however, was through a multiplicity of informal relationships among them built initially around the science park in Hsinchu. Graduates of the four engineering universities certainly produced an elite cohort of intersecting relationships across the large companies that came to be known as the Tai Cheng Tsing Chiao network. Beyond them, a huge array of small companies emerged in the region to fill gaps and niches through myriad practices of production and contracting (D. Chen 2008; Dedrick and Kraemer 1998; Shieh 1992).

The rapid rise of the electronics industry helped to materialize the islands of Taiwan, Penghu, Kinmen, and Matsu into a territory distinct from the mainland (Harrison 2006). Loss of U.S. recognition contributed to this process by asserting that the Republic of China on these islands could no longer claim hegemony over the mainland. The government's agreement and turn to support private industry on the land of Taiwan produced a dramatic step, acknowledging that the main island was becoming more than a temporary home for the mainlanders who had fled.

The new development of electronics on Taiwan thus also became the development of electronics for Taiwan. The arena of private entrepreneurial initiative that the KMT government had long ignored at a distance became an arena of interest to the government (Y. Wu 2005). But electronics technologies resisted the straightforward transfer of practices from the chemical, sugar, and other industries that were built on hierarchical factories and were dominated by state-owned and state-supported companies. Mainlanders could not maintain control of the development of electronics through direct supervision or subcontracting with small islander companies for parts and services around the margins. Government-owned foundries for semiconductors were certainly essential to all sectors of the industry but could not serve as its center. Key centers in electronics manufacturing would not be stable in the case of Taiwan.

Advocates for the new industry like to document its benefits to Taiwan by calling attention to increases in income and gross national product (GNP). Certainly, between 1952 and 1992, the average personal income in Taiwan increased from $200 to $10,000 USD, with most of these increases taking place during the period 1968 to 1988. Prices remained relatively stable and the distribution of income remained relatively equitable when compared with other countries of the world.[7] The GNP increased from $5 billion in

[7] The ratio of income of the top quintile to the lowest quintile declined from 5.28 in 1968 to 4.7 in 1987. The Gini concentration coefficient declined from 0.362 in 1968 to 0.307 in 1987, among the best in the world.

1969 to $105 billion in 1987 (National Statistics 2023). The rise in domestic production slowed and reversed the flow of international companies onto the island as local companies supplanted international firms (Amsden and Chu 2003). At the same time, increased income had different implications for the two main geographical identities.

For mainlanders, the emergence of a distinctive transnational identity as an electronics powerhouse could be linked to that of the Republic of China. It could support further development of the KMT vision of Taiwan as host for Free China. The many technical contributions of Hu Ding-Hua and other mainlander engineers appeared to aim in that direction (Hu 2011; S. Wu 2019).

Yet the work of Chew Tsai-Hsin and other islander engineers and engineering technicians did not easily fit that vision. Private-sector industries had largely been a sidelight for even the government of Chiang Ching-Kuo, just as they had long been for Chiang Kai-Shek. After the "War of Resistance" (Chiang and Jaffe 1947 (1943), 133–148), the marginal status of private industry had freed islanders to look for work, build careers, and so on, but mostly in small companies that were labor-intensive (Lin 2009). By now pursuing and embracing careers in an industry that was producing Taiwan as a thing on the world stage, Chew and other local engineers were also contributing materially to the emergence of something different than another version of China led by mainlanders. Minimally, it was an image of Taiwan as a thing unto itself, separate from the mainland rather than attached to it.

Hu Ding-Hua, Chew Tsai-Hsin, and the multiplying engineers and engineering technicians were carrying technical education in and around electrical engineering into technical work designing, building, and maintaining industrial practices in electronics. In at least one key sense, that technical work also carried an important nontechnical meaning and significance for watchers and listeners both inside and outside of Taiwan—it called attention to Taiwan as Taiwan. At a time when the United States was ceasing foreign aid to the Republic of China and building a new relationship with the government of the People's Republic of China, calling attention to Taiwan as Taiwan enacted a material commitment that was at once both profound and courageous.

Years later, the second-generation mainlander Terry Guo, who founded one of the largest electronics manufacturers, was justifying to an official from Singapore the KMT government's early support for large companies, forcing small companies to fend for themselves. He asserted that supporting the small companies would have hurt them, like "if you gave a sugar to a three-year-old baby that was hard to feed." They would not have become "so competitive" (Guo 2004, 88–89).

Whether or not he was right, Guo likely did not realize that he was also making a very different point—that the small and, then, medium sized companies built and nurtured by islander families deserved acknowledgement as essential partners alongside mainlanders in the electronics industry. In an important sense, the rise of electronics had helped to turn all engineers and engineering technicians in Taiwan into islanders. Perhaps that was the real miracle of the electronics industry – as electronics engineers and technicians animated two Taiwans, the "abnormal club" got a lot bigger.

It is important to recognize that the KMT version of growth and success for Taiwan hid many continuing inequities. Hierarchical power relations separating mainlanders from islanders (and both from indigenous peoples) did not disappear. However, by helping to legitimize a version of Taiwan as a thing apart from the mainland, electronics engineers and technicians made it increasingly difficult for the KMT government to avoid such issues. By animating two Taiwans, they provided material grounds for the democratization movements that would emerge during the next decade.

References

Amsden, Alice H., and Wan-wen Chu. 2003. *Beyond Late Development: Taiwan's' Upgrating Policies*. Cambridge, MA: MIT Press.

Anonymous. 1956. "Ji nian da hui ji lu 紀念大會紀錄 (The Commemoration Minutes)." *Jiao da you sheng 交大友聲 (The Alumni Voice of Chiao Tung University)* 42: 9–12.

Anonymous. 1958. "New York jiao da xiao you hui zhi xing wei yuan hui hui yi ji lu 紐約交大校友會執行委員會會議紀錄 (The Meeting Minutes of Executive Committee of the Chiao Tung Alumni Association of New York)." *Jiao da you sheng 交大友聲 (The Alumni Voice of Chiao Tung University)* 62: 41–42.

Anonymous. 1966. "Dian zi yan jiu suo jin xun 電子研究所近訊 (News from the Institute of Electronics)." *Jiao da you sheng 交大友聲 (The Alumni Voice of Chiao Tung University)* 163: 32–39.

Chang, Chun-yen 張俊彥. 2011. "Taiwanese IT Pioneers: Chun-yen Chang." Interviewed by Craig Addison. Mountain View, CA: Computer History Museum.

Chang, Chun-yen 張俊彥, and Po-Lung Yu 游伯龍, eds. 2001. *Made by Taiwan: Booming in the Information Technology Era*. River Edge, NJ: World Scientific.

Chang, Ju-Hsin 張如心, and Pan Wen Yuan Foundation 潘文淵文教基金會. 2006. *Xi shuo tai wan: Tai wan ban dao ti chan ye chuan qi 矽說台灣: 台灣半導體產業傳奇 (Silicon Taiwan: The Legend of Taiwan's Semiconductor Industry)*. Taipei: Commonwealth Publishing Group (天下文化).

Chang, Kuo-Hui, Gary Lee Downey, and Bono Po-Jen Shih. 2020. "Talents First!: Wu Ta-you and Science Policy Infrastructures in the Republic of China (1927–1970)." *The Korean Journal for the History of Science* 42(2): 449–473.

Chen, Dung-Sheng 陳東升. 2008. *Ji ti wang lu: Tai wan gao ke ji chan ye de she hui xue fen xi 積體網絡: 台灣高科技產業的社會學分析 (Organizational Networks in Taiwan's Integrated-circuit Industry)*. Taipei: Socio Publishing (群學).

Chen, Shun-Fen 陳舜芬. 1991. "Guang fu hou tai wan di qu gao deng jiao yu she xiao zheng ce zhi tan tao 光復後台灣地區高等教育設校政策之探討 (Discussion on the Policies of Establishing Schools of Higher Education in Taiwan after the Restoration)." In *Guang fu hou taiwan fa zhan jing yan 光復後台灣發展經驗 (Taiwan's Development Experience after the Restoration)* Vol. 27, edited by Jeh-Hang Lai 賴澤涵 and Chun-chieh Huang 黃俊傑, 205–237. Taipei: Academia Sinica.

Cheng, Kuo-Ping 鄭國賓. 1993. "From METS to TAC." In *Wo suo ren shi de sun yun xuan: Sun yun xuan ba shi da shou ji nian zhuan ji 我所認識的孫運璿: 孫運璿八十大壽紀念專輯 (What*

I Know about Sun Yun-suan: Memorial Collection for Sun Yun-suan Eighty's Birthday) edited by Sho-Tzu Chiu 丘秀芷, 170–176. Taipei: Sun Yun-suan Foundation.

Zheng, Youkui 郑友揆, Cheng Linsun 程麟荪, and Chuanhong Zhang 张传洪. 1991. *Jiu zhong guo de zi yuan wei yuan hui (1932–1949): Shi shi yu ping jia* 旧中国的资源委员会 *(1932–1949):* 史实与评价 *(National Resources Commission in Republican China, 1932–1949: Historical Facts and Evaluation)*. Shanghai: Shanghai Academy of Social Sciences Press.

Chiang, Kai-shek, and Philip J. Jaffe. 1947 (1943). *China's Destiny and Chinese Economic Theory*. New York: Roy Publishers.

Chiu, Tsai-Hsin 邱再興. 2015. *She de: Dian zi ye xian qu qiu zai xing de shi ye yu zhi ye* 捨得*:* 電子業先驅邱再興的事業與志業 *(Willing to: Career and Aspirations of Electronics Pioneer Chew Tsai-hsing)*. Taipei: Booklife (圓神).

Dedrick, Jason, and Kenneth L. Kraemer. 1998. *Asia's Computer Challenge: Threat or Opportunity for the United States and the World?* New York: Oxford University Press.

Department of Electronics Engineering, National Chiao Tung University. 2008. *Jiao da dian zi suo 50 zhou nian ji nian DVD: Te bie sou cang* 交大電子所*50*周年紀念*DVD:* 特別蒐藏 *(Memorial DVD of 50 Years of Department of Electronics Engineering, National Chiao Tung University: Special Video Collections)*. Hsinchu, Taiwan: Department of Electronics Engineering, National Chiao Tung University.

Department of Statistics, Ministry of Education, Taiwan 教育部統計處. 1990. *Zhong hua min guo da zhuan yuan xiao li jie bi ye ren shu tong ji* 中華民國大專院校歷屆畢業人數統計 *(Statistics of the Number of Graduates in Tertiary Education of R.O.C.)*. Taipei: Ministry of Education.

Greene, J. Megan. 2008. *The Origins of the Developmental State in Taiwan: Science Policy and the Quest for Modernization*. Cambridge, MA: Harvard University Press.

Guo, Terry Tai-Ming 郭台銘. 2004. "Guo tai ming de chuang xin jing ying 郭台銘的創新經營 (Terry Guo's Innovative Management)." In *Biao gan ling xiu jiao ni jing ying zhe xue* 標竿領袖教你經營哲學*:* 九位最有影響力台灣企業家的變革、經營哲學 *(Leader to Leader: Nine of Most Inflential Taiwanese Entreprenuers' Innovative Wisdom of Change)*, edited by Commonwealth Magazine, 81–89. Taipei: Commonwealth Publishing Group (天下文化).

Han, Kyonghee, and Gary Lee Downey. 2014. *Engineers for Korea*. Cham, Switzerland: Springer International Publishing.

Harrison, Mark. 2006. *Legitimacy, Meaning and Knowledge in the Making of Taiwanese Identity*. New York: Palgrave Macmillan.

Ho, Ming-sho 何明修. 2016. *Zhi li po sui de tuan jie: Zhan hou tai wan lian you chang yu tang chang de lao gong* 支離破碎的團結：戰後台灣煉油廠與糖廠的勞工 *(A Fragmented Unity: Workers in Taiwan's Oil Refinery and Sugar Factory in the Postwar Period)*. New Taipei City: Rive Gauche (左岸).

Ho, Pang-Li 何邦立, ed. 2004. *He yi ci xian sheng ji* 何宜慈先生集 *(Irving T. Ho Memorial Collection)*. Taipei: Irving T. Ho Memorial Foundation.

Hong, Sao-Yang 洪紹洋. 2016. "1950 nian dai tai ri jing ji guan xi de zhong qi yu tiao zheng 1950年代臺、日經濟關係的重啟與調整 (Re-establishment and Adjustment of Taiwan-Japan Economic Relations in 1950s)." *Tai wan shi yan jiu* 臺灣史研究 *(Taiwan Historical Research)* 23(2): 165–210.

Hou, Chi-ming, and San Gee. 1993. "National Systems Supporting Technical Advance in Industry: the Case of Taiwan." In *National Innovation Systems: A Comparative Analysis*, edited by Richard R. Nelson, 384–413. New York: Oxford University Press.

Hu, Ding-Hua 胡定華. 2011. "Taiwanese IT Pioneers: Ding-Hua Hu." Interviewed by Ling-Fei Lin 林玲妃. Mountain View, CA: Computer History Museum.

Huang, Jinlin 黄金麟. 2017. "Zhan zheng yu taiwan de gao deng jiao yu, 1945–1990 戰爭與台灣的高等教育, 1945–1990 (War and Higher Education in Taiwan, 1945–1990)." *Tai wan she hui xue* 台灣社會學 *(Taiwanese Sociology)* 34: 123–167.

Industrial Technology Research Institute 工業技術研究院. 2013. *Gong yan yuan si shi nian da shi ji* 工研院四十年大事紀 *(40 Years of Industrial Technology Research Institute)*. Hsinchu, Taiwan: Industrial Technology Research Institute.

Jacoby, Neil H. 1967. *U.S. Aid to Taiwan: a Study of Foreign Aid, Self-help, and Development*. New York: F. A. Praeger.

Johnstone, Bob. 1994. "Me-Too Is Not My Style." *Wired* 2(9): 125–129.

Kao, Shu-Yuan 高淑媛. 2016. *Tai wan gong ye shi* 臺灣工業史 *(Taiwan Industrial History)*. Taipei: Wunan (五南).

Kirby, William C. 1990. "Continuity and Change in Modern China: Chinese Economic Planning on the Mainland and on Taiwan, 1943–1958." *Australian Journal of Chinese Affairs* 24: 121–141.

Lam, Barry Pak-lee 林百里. 2011. "Taiwanese IT Pioneers: Barry (Pak-Lee) Lam." Interviewed by Ling-Fei Lin 林玲妃. Mountain View, CA: Computer History Museum.

Lin, Ching-Po 林清波, ed. 2012. *Tai wan ying zao ye bai nian shi* 臺灣營造業百年史 *(Taiwan's Construction Industry: A Centenary History)*. Taipei: Yuan-Liou (遠流).

Lin, Chung-Hsi. 2009. "The Silenced Technology—the Beauty and Sorrow of Reassembled Cars." *East Asian Science Technology and Society: An International Journal* 3(1): 91–131.

Lin, Hsiao-ting 林孝庭. 2021. *Jiang jing guo de tai wan shi dai* 蔣經國的台灣時代:: 中華民國與冷戰下的台灣 *(The Chiang Ching-Kuo Era: The Republic of China on Taiwan in the Cold War)*. Taipei: Walkers Cultural Enterprises (遠足).

Lin, Yi-Ping, and Hsin-Hsing Chen. 2021. "STS on the Street and at the Court: Interlocutors in the Taiwan RCA Collective Occupational Disease Lawsuit." In *Making and Doing: Activating STS through Knowledge Expression and Trave*, edited by Gary. L. Downey and Teun Zuiderent-Jerak, 37–50. Cambridge, MA: MIT Press.

Liu, Shinkei 劉進慶. 2012 [1992]. *Tai wan zhan hou jing ji fen xi* 台灣戰後經濟分析 *(Taiwan Postwar Economic Analysis)*. Taipei: Renjien (人間).

Liu, Shu-Fen 劉素芬. 2005. *Li guo ding: Wo de tai wan jing yan* 李國鼎: 我的台灣經驗 *(Kwoh-Ting Li: My Taiwan Experience)*. Taipei: Yuan-Liou (遠流).

McCusker, Henry. F., and Harry L. Robinson. 1962. *Education and Development: The Role of Educational Planning in the Economic Development of the Republic of China*. Menlo Park, CA: Stanford Research Institute.

Ministry of Education, Taiwan 教育部. 1971. *Zhong hua min guo jiao yu tong ji* 中華民國教育統計 *(Educational Statistics of the Republic of China, 1970)*. Taipei: Ministry of Education.

National Statistics, Taiwan 國情統計. 2023. Guo min suo de ji jing ji cheng zhang tong ji zi liao ku 國民所得及經濟成長統計資料庫 (National Income and Economic Growth Statistics Database). Taipei: DGBAS (行政院主計處).

Otellini, Paul S. 2006. "Stan Shih." *Time*, Nov. 13.

Pan Wen Yuan Foundation. 2006. "Pan wen yuan ji jin hui di si qi hui kan 潘文淵基金會第四期會刊: 十周年紀念特刊 (The 4th Report of Pan Wen Yuan Foundation: Special Issue of the Tenth Anniversary)." *Pan wen yuan ji jin hui hui kan* 潘文淵基金會會刊 *(Pan Wen Yuan Foundation)*, 43–46. Hsinchu, Taiwan. Pan Wen Yuan Foundation.

Pan, Wen Yuan 潘文淵. 1974. "Letter to Sun Yun-Suan on the Development of Taiwan's Electronic Integrated Circuit Technology Plan." Retrieved from https://iht.nstm.gov.tw/industry/index-1.asp?m=&m1=3&m2=15&id=63.

Pan, Wen Yuan 潘文淵. 1985. "The Founding of METS." In *Jin dai gong cheng ji shu tao lun hui shi jie ji nian zhuan ji: min guo wu shi wu nian-min guo qi shi si nian* 近代工程技術討論會十屆紀念專集: 民國五十五年-民國七十四年 *(Special Memorial Volume of Ten Modern Engineering*

Technology Seminars), edited by Yen-Shiang Shih 施顏祥, 43–46. Taipei: Chinese Institute of Engineers.

Saxenian, AnnaLee. 2006. *The New Argonauts: Regional Advantage in a Global Economy*. Cambridge, MA: Harvard University Press.

Shieh, Gwo-Shyong. 1992. *"Boss" Island: The Subcontracting Network and Mico-Entrepreneurship in Taiwan's Development*. New York: Peter Lang.

Shih, Stan. 1996. *Me-too Is Not My Style*. Taipei: Acer Foundation.

Shih, Stan. 2011. "Taiwanese IT Pioneers: Stan (Chen-Jung) Shih." Interviewed by Ling-Fei Lin 林玲妃. Mountain View, CA: Computer History Museum.

Shih, Stan. 2018. *Hong qi jing yan yu tai wan dian zi ye: Shi zhen rong xian sheng fang wen ji lu 宏碁經驗與台灣電子業: 施振榮先生訪問紀錄 (Acer's Experience and Taiwan's Electronics Industry: The Reminiscences of Stan Shih)*. Interviewed by Miao-Fen Lu 呂妙芬 et al. Taipei: Institute of Modern History, Academia Sinica.

Shih, Yen-Shiang 施顏祥, ed. 1985. *Jin dai gong cheng ji shu tao lun hui shi jie ji nian zhuan ji 近代工程技術討論會十屆紀念專集 (Special Memorial Volume of Ten Modern Engineering Technology Seminars)*. Taipei: Chinese Institute of Engineers.

Taylor, Jay. 2000. *The Generalissimo's Son: Chiang Ching-kuo and the Revolutions in China and Taiwan*. Cambridge, MA: Harvard University Press.

Tsao, Richard Hsin-Chen 曹興誠. 1999. *Cao xing cheng: Lian dian de ba ye chuan qi 曹興誠: 聯電的霸業傳奇 (Robert Tsao: The Legend of UMC's Dominance)*. Taipei: Commonwealth Publishing Group (天下文化).

Tsao, Richard Hsin-Chen 曹興誠. 2004. Cao xing cheng de bian ge zhi hui 曹興誠的變革智慧 (Robert Tsao's Wisdom of Change). In *Biao gan ling xiu jiao ni jing ying zhe xue 標竿領袖教你經營哲學: 九位最有影響力台灣企業家的變革、經營哲學 (Leader to Leader: Nine of Most Inflential Taiwanese Entreprenuers' Innovative Wisdom of Change)*, edited by Commonwealth Magazine, 59–78. Taipei: Commonwealth Publishing Group (天下文化).

Tsao, Tseng-Chueh 趙曾玨. 1964. "My Impression of Taiwan: The Free China." *Jiao da you sheng 交大友聲 (The Alumni Voice of Chiao Tung University)* 144: 3–9.

Tsao, Tseng-Chueh 趙曾玨. 1985. "The Impacts of METS." In *Jin dai gong cheng ji shu tao lun hui shi jie ji nian zhuan ji 近代工程技術討論會十屆紀念專集: 民國五十五年-民國七十四年 (Special Memorial Volume of Ten Modern Engineering Technology Seminars: 1966–1975)*, edited by, Yen-Shiang Shih 施顏祥, 55–62. Taipei: Chinese Institute of Engineers.

Tsao, Tseng-Chueh 趙曾玨, and David K. Cheng 鄭鈞. 1964. "Jin nian xiao qing de zui jia shou li 今年校慶的最佳壽禮 (The Best Longevity Birthday Gift for This Year's School Anniversary)." *Jiao da zhi sheng 交大友聲 (The Alumni Voice of Chiao Tung University)* 136: 1–8.

UNESCO, United Nationas Educational Scientific and Cultural Organization. 1964. *Economic and Social Aspects of Educational Planning*. Paris: UNESCO.

Wang, Hsi-Chi 王仕琦. 2012. *Fu zi shuang jie, qing hua chuan cheng 父子雙傑清華傳承 (Outstanding Father and Son in Tsing Hua)*. Hsinchu, Taiwan: National Tsing Hua University.

Wang, Jing 王璟. 1985. "Jin dai gong cheng ji shu tao lun hui yuan qi ji fa zhan jing guo 近代工程技術討論會緣起及發展經過 (The Origin and Development Process of Modern Engineering Technology Seminar)." In *Jin dai gong cheng ji shu tao lun hui shi jie ji nian zhuan ji 近代工程技術討論會十屆紀念專集 (Commemorative Collection of the Tenth Symposium on Modern Engineering Technology Seminar)*, edited by Yen-Shiang Shih 施顏祥, 47–53. Taipei: Chinese Institute of Engineers.

Wang, Li-Chun 王麗娟. 2013. *Shi min yu shu wei shi dai de gu shi 施敏與數位時代的故事 (Simon M. Sze and the Stories of the Digital Era)*. Taipei: Wa-People (紅津).

Wu, Su-Min 吳淑敏. 2019. *Hu ding hua chuang xin xing zhuan 胡定華創新行傳 (Hu Ding-Hua and the Acts of His Innovation)*. Hsinchu, Taiwan: Rehoboth Pioneer Networking (力和博).

Wu, Tsong-Min. 吳聰敏. 1988. "Mei yuan yu tai wan de jing ji fa zhan 美援與台灣的經濟發展 (US Aid and Taiwan's Economic Development)." *Tai wan she hui yan jiu ji kan*台灣社會研究季刊 *(Taiwan: A Radical Quarterly in Social Studies)* 1: 145–158.

Wu, Tsong-Min 吳聰敏. 2023. *Tai wan jing ji si bai nian* 台灣經濟四百年 *(A 400-Year History of Taiwanese Economy)*. Taipei: Spring Hill (春山).

Wu, Ta-you 吳大猷. 1989. "Jiang gong yu wo guo ke xue fa zhan 蔣公與我國科學發展: 命我就科學發展指導委員會職的經過 (Chiang Kai-shek and Science Development in Our Country: How I Became the Head of the Advisory Committee for Science Development)." In *Zai tai gongzuo huiyi* 在台工作回憶 *(Memories of Works in Taiwan)*, edited by Ta-you Wu, 68–70. Taipei: Yuan-liou (遠流).

Wu, Yongping. 2005. *A Political Explanation of Economic Growth: State Survival, Bureaucratic Politics, and Private Enterprises in the Making of Taiwan's Economy, 1950–1985*. Cambridge, MA: Harvard University Press.

Yang, Ai-Li 楊艾俐. 1989. *Sun yun hsuan chuan* 孫運璿傳 *(Biography of Sun Yun-Suan)*. Taipei: Commonwealth Publishing Group (天下文化).

Yang, Ding-Yuan 楊丁元. 2011. "Taiwanese IT Pioneers: D. Y. (Ding-Yuan) Yang." Interviewed by Ling-Fei Lin 林玲妃. Mountain View, CA: Computer History Museum.

Yang, Ding-Yuan 楊丁元, and Hui-Ling Chen 陳慧玲. 1996. *Ye jing tian ze: Gao ke ji chan ye sheng tai* 業競天擇: 高科技產業生態 *(Ecology of Taiwan's High-tech Industry)*. Taipei: Commercial Times (工商時報).

5 Two Taiwans in the U.S.

It finally happened in 1991. Islander engineers living and working in the United States established a professional association for engineers that included the island in its name. The founding president, Yu John Pang (余金榜, 1937–), explained that he and other islanders in Silicon Valley…

> ... had been thinking a lot [that] one day .... Taiwanese engineers and scientists would want to be recognized in international events that they are from Taiwan. But we actually did not use the name of Taiwan. We would be introduced to foreigners that we are from China. They would think we are Chinese.[1]

Yu had come to the United States in 1965 to study electrical engineering after finishing an associate degree at Provincial Taipei Institute of Technology (hereafter Taipei Institute, now known as National Taipei University of Technology), a junior college attended mostly by islanders. He completed both bachelor and master's degrees at Kansas State University, followed by a Ph.D. at the University of Connecticut. He then spent a decade designing and developing consumer electronics at Bell Labs before setting off to manage new engineering departments for other companies in Silicon Valley in the mid-1980s.

Establishing the new association was both an achievement and a relief for many islander engineers. It had become possible only after Lee Teng-Hui (1923–2020) became president of the Republic of China in 1988, following the lifting of martial law the previous year by Chiang Ching-Kuo (1910–1988). Lee was both the first president and the first chairman of the Kuomintang Party who was born in Taiwan. Yet the new association did not actually use the name "Taiwan." Rather it referred to itself as a collection of members called "Taiwanese": North American Taiwanese Engineers' Association (later renamed North America Taiwanese Engineering & Science Association, or NATEA).

[1] Yu John Pang, interviewed by Chang Kuo-Hui, April 14, 2018.

K.-H. Chang et al., *Engineers and the Two Taiwans*, Synthesis Lectures on Global Engineering, https://doi.org/10.1007/978-3-031-59766-4_5

As we shall see below, the distinction between "Taiwan" and "Taiwanese" included a significant difference.

The episode we followed in Chap. 4 showed how engineers and engineering technicians in the Republic of China helped to animate two versions of Taiwan from the late 1960s to the early 1980s through their technical work in electronics industries. While this was taking place, engineers who traveled to the United States for education and work were doing the same. It's just that they animated two Taiwans much more explicitly and in a wholly different way—by creating, naming, and joining professional associations. For engineers who went to the United States and found jobs with U.S.-based companies, their career transitions necessarily involved adding new identities associated with that country. However, those transitions did not require disavowing or ignoring other identities that those engineers had previously established back home, whether home was somewhere on the mainland, in Taiwan, or a combination of both.

The episode in this chapter follows how mainlander and islander engineers developed and positioned professional associations in the United States. Would-be associations faced difficult, evolving challenges to define and distinguish their members not only from engineers in the People's Republic of China but also from one another. For those who traveled to the United States, the pursuit of engineering learning, engineering work, and everyday life there prompted individual assessments of career attachments, life attachments, and relationships between them. Joining a professional association that identified itself with Taiwan and sought to serve Taiwan in particular ways tended to make those calculations quite explicit. Still, for many engineers and scientists, likely a majority, serving Taiwan appropriately meant avoiding any identification with such associations. Joining meant embracing that kind of politics that engineers tend to resist—partisan politics.

The "C" Camp

Mainlander engineers traveled to the United States first. During the 1950s and 1960s, it was common among mainlanders to view Taiwan as a remote, rural, i.e., pre-industrial, province that was but a temporary refuge for the KMT government. Many engineers who had fled the mainland redirected themselves to the United States to build careers and, possibly, stay there forever.

As one early graduate of National Chiao Tung University put it, "many KMT officials sent their children out" to the United States to "keep a seed." "In case Taiwan is gone [because the Communists invaded] and the whole family is gone," he explained, "at least families still had one planted in the United States." This observation came from an islander, Ku Po-Kwan (顧博光, 1941–), who elected to stay in Taiwan and devote his career to teaching, mostly at Feng Chia University. Musing about the many mainlander students who left, Ku remembered that "that group of people never regarded Taiwan as their homeland, nor did they identify with Taiwan." Rather, "they … identified with the mainland," likening temporary living in Taiwan to floating like "duckweed."[2]

By the 1970s and 1980s, as the nationalist hope of retaking the mainland was fading, mainlander engineers who left regularly took their whole families with them, including grandparents. An electrical engineering graduate from National Taiwan University in 1976 reported that 90 of the 110 graduates in his class went "abroad" for further degrees, with "abroad" mostly code for the United States. "Most of my classmates with mainlander background," he said, "hoped they would find a nice job staying in the U.S. and took their original family to live with them."[3]

We saw in Chap. 4 that the many mainlander engineers who helped the KMT government found or support electronics companies in Taiwan during the 1960s and 1970s were often members of the Chinese Institute of Engineers (hereafter Chinese Institute). The Chinese Institute had been founded in New York in 1917 and then relocated to Taiwan in 1951.[4] While older members in the 1950s and 1960s may have still held out hope that the KMT government would re-take the mainland, younger members often enthusiastically embraced the government's emerging commitment to industrial development on the island of Taiwan itself. It was younger members who mobilized the Chinese Institute to support the Modern Engineering and Technology Seminar, which to this day meets biennially to share technical knowledge and visions with engineers and government officials in the Republic of China.[5]

Mainlander engineers also joined other professional associations in the United States, especially the broadly-based Chinese-American Professionals Association (國建聯誼會, hereafter Professionals Association). Established in Washington, D.C. in 1975 with financial support from the KMT government, the Professionals Association sought to attract technical experts whose capabilities extended beyond the Chinese Institute's focus on practicing industrial technologists (Chao 2007). As the organization's website still reports, its members "carried out important advisory roles, contributing to the construction and development of the Republic of China (ROC)."[6] Even as some members found research positions in the executive branch of the U.S. government or in contractors for the U.S. government, the work of advising the KMT government re-affirmed attachments to the Republic of China.

---

[2] Ku Po-Kwan, interviewed by Chang Kuo-Hui, January 5, 2020. Another islander engineer reported confidentially that "parents of my classmates from other provinces … absolutely wanted them to stay [in the United States] because they were afraid of the Communist Party."

[3] Roger Liao (廖如圭, 1954–), interviewed by Chang Kuo-Hui, April 3, 2018.

[4] The Chinese Institute also later established chapters in such U.S. cities as San Francisco and Detroit, mostly for mainlander civil engineers. Chiu Chun-Pang, interviewed by Chang Kuo-Hui, April 13, 2018.

[5] See http://www.cie.org.tw/English, accessed on October 16, 2023.

[6] See https://sites.google.com/view/capadc/home, accessed on October 16, 2023.

Some mainlander engineers who were building careers in and around teaching and research joined the Chinese American Academic and Professional Society (美東華人學術聯誼會, hereafter Academic and Professional Society), which was also established in New York in 1975. In addition to promoting scholarly exchanges and fellowship among Chinese-American researchers and teachers, i.e., persons of Han descent and nationalist orientation, the Academic and Professional Society sought to promote and lead Chinese Americans generally in relation to other ethnic groups in the United States. It pictured itself and them as merging into an American melting pot. Years later, for example, the theme of the 29th annual convention of the Academic and Professional Society in 2004 explicitly championed its longtime goal of integrating its members in the new country: "Establish, Grow, and Merge into the Main Stream in America."[7]

Important for our purposes is that all these organizations used the "C" word in their names: Chinese. Naming performs a kind of politics, in the sense of defining power relations among distinct entities. Just as an act of defining an engineering problem in mathematical terms attaches one's disciplinary identity to it (Bucciarelli 1994), so did naming all these professional organizations "Chinese" attach to them a KMT vision of Taiwan as "Free China." It was certainly the case that all openly welcomed as members islander engineers who had traveled to the U.S. However, membership for islanders came with a familiar catch. In the same way that the new KMT initiatives in Taiwan to build and support large companies took for granted and expected to reproduce a hierarchy of mainlanders over participating islanders, so did the Chinese associations expect that mainlanders would lead and non-mainlanders would follow.

Consider, for example, the islander engineer Huang Kuan-Tsae (黃光彩, 1953–) who first joined the Taiwanese Association of America. Huang arrived in the U.S. in 1978 after completing a degree in mathematics at National Taiwan Normal University (hereafter Taiwan Normal), a teachers' college primarily for islanders (Su and Yu 2007, 33–34). After concluding a master's in mathematics at the University of Illinois and while working toward a first Ph.D. at MIT in electrical engineering/computer science, Huang noted a contrast among students attending graduate school in Cambridge, Massachusetts. Those enrolled at MIT tended to study engineering, including both islanders and mainlanders, while those at Harvard tended to be mainlanders. Huang and his islander colleagues at MIT had concerns about a particular mainlander among them observing and recording their actions. "We worried," Huang said, that he might "report us to the KMT government."[8]

After starting work at IBM in 1982, Huang then joined the Academic Society in New York. He quickly learned that "islanders had a hard time having voices and stepping up to lead." During the 1980s, Huang encountered numerous other islander engineers who,

[7] See https://www.caaps.us/2004Activities.html, accessed on October 16, 2023.

[8] Huang Kuan-Tsae, interviewed by Chang Kuo-Hui, March 31, 2018.

like him, had "received invitations" to join those organizations but "felt … that they were not accepted to lead or manage."[9]

Another early islander engineer who had graduated from Taiwan Normal, this time in physics, was Chi Gou-Chung (紀國鐘, 1946–). After serving in the military and teaching secondary school in Taiwan, Chi successfully gained admission to Yale University. He completed a Ph.D. in engineering and applied science in 1976 and then initiated a long career at Bell Labs. When the Professionals Association reached out to him to join, he declined. "I felt," he said, that the islanders in that organization "were its decoration in terms of political representation."[10]

The "F" to "T" Camp

In contrast with mainlanders, early islander engineers who came to the United States tended not to bring their families. For one thing, "the cost of the air ticket was very … expensive at that time," explained Chew Tsai-Hsin, the Abnormal Club islander we followed in Chap. 4. While "it was not difficult to get a scholarship" in the United States, "that ticket was difficult," so "it was not affordable for ordinary people," i.e., islanders.[11]

There was also an especially difficult issue of identity for islander engineers who traveled to the United States—were they detaching themselves from Taiwan? The graduate above from National Taiwan University who described mainlanders as taking their families also explained that his islander classmates usually came alone. They "did not," he said, "want to live the rest of their lives in the U.S." In Taiwan, they had typically grown up in a nuclear family, and the land itself held the graves of generations of family members who preceded them.[12] The land could not be transported to the United States.

The Taiwan Normal graduate who went to work at Bell Labs, Chi Gou-Chung, explained that early islanders learning and working in the United States tended to see it as temporary, a kind of "paradise" that afforded freedoms unavailable at home. At the same time, having families still in Taiwan "put a lot of cultural pressure on us," he said, meaning that responsibilities there could not simply be ignored or put on hold. "For example," he elaborated, "you want to sweep the graves during the Qingming Festival," which was observed both on the mainland and in Taiwan as Ancestors' Day or Tomb-Sweeping

[9] Huang Kuan-Tsae, interviewed by Chang Kuo-Hui, March 31, 2018. See also (Y. Chen 2009b, 397–398).

[10] Chi Gou-Chung, interviewed by Chang Kuo-Hui, April 10, 2018.

[11] Chew Tsai-Hsin, interviewed by Chang Kuo-Hui, January 29, 2021. Also see Y. Chen (2009b, 326).

[12] Roger Liao, interviewed by Chang Kuo-Hui, April 3, 2018.

Day. It was not surprising, then, that many islander engineers took care to pay regular visits to Taiwan.[13]

As early as the 1960s, some islander engineers who came to the United States publicly resisted the KMT government, thereby accepting a politics of difference or conflict. The popular membership association, United Formosans in America for Independence (United Formosans) was the most explicit about this goal. Founded in 1966, United Formosans understood independence as terminating the KMT government occupation of Taiwan and formalizing the island's separation from the mainland. Using the name "Formosans" highlighted an attachment to the main island without reference to China.

Through the 1960s and 1970s, organizers of islander associations tended to prefer "Formosa" to "Taiwan." The key reason was that the Chinese original for Taiwan—臺灣—was just that, a Chinese name born on the mainland. Formosa was also preferable because it was recognized internationally, albeit through a perspective that was clearly and explicitly Western (Andrade 2008). The association of "Taiwan" with China was more important to resist at that time than memory of an attachment to Europe, especially for an association founded in the United States. Other related associations in the U.S. included the West Coast Formosa Club (f. 1960) and the Formosa Association for Public Affairs (f. 1982).

Despite the fact that United Formosans encouraged and attracted membership from all islanders in the U.S., engineers became its core. Over its first 30 years of existence, for example, seven of its 13 chairmen were engineers while the other six were distributed across professional fields (Chang and Tseng 2009, 80). Chi Gou-Chung was one of the early activist organizers on behalf of islander engineers. He reported that he and others who received their early training in Taiwan were wholly unaware that United Formosans even existed in the United States before going there themselves. And being able to join any association was itself a statement of separation from the KMT government. Because of martial law, he said, "our generation was not allowed to assemble and form associations in Taiwan."[14] The freedom to associate was one marker of the paradise that came with study and work in the United States.

A key locus of engineers organizing for independence was Kansas State University. In 1961, a group of islander graduate students in engineering founded the first Taiwan Student Association in the United States. It took its place alongside the existing mainlander organization, the Chinese Student Association, before the name "Formosa" became a preferable alternative. Those students also joined with islander professors to organize the K-State Group, an informal gathering of students and professors that regularly discussed political issues in Taiwan. Some engineering professors served as prominent participants,

[13] Chi Gou-Chung, interviewed by Chang Kuo-Hui, April 10, 2018. Some of the senior Taiwanese engineers who had worked for such U.S.-based companies as Intel, Cisco and AMD offered similar responses. Rockwell Hsu, Yeh Chiunn-Shyong (Eugene) and Roger Liao, interviewed by Chang Kuo-Hui, April 22, 2018, April 14, 2018 and April 3, 2018.

[14] Chi Gou-Chung, interviewed by Chang Kuo-Hui, April 10, 2018.

including Fan Liang-Tseng (范良政, 1929–2014), professor of chemical engineering and brother of another member who was professor of economics (Fan Liang-Shing 范良信, 1932–2016).[15]

Among islander professionals, Kansas State became known as "West Point for independence activists," a reference to the Republic of China Military Academy that was founded in 1924 to be the West Point of the East (M. Chen 1992, 103; Y. Chen 2009b, 329, 393). Yang Chung-Chang (楊宗昌; 1936–2023), for example, arrived at K-State in 1962 to study civil engineering after completing his associate degree at Taipei Institute. During his second year, a Ph.D. student in electrical engineering introduced him to United Formosans, and he developed into a key organizer. The then-president of the student association Strong Chuang (莊秋雄,1939–2022), who was pursuing a master's degree of engineering, remembered that Yang

> designed a complicated contact method for [United Formosan] that prevented them being caught.… [W]hen we made phone calls, he made clear rules of what we should ask and how we should respond. In addition, all of the phone numbers we shared were secretly coded and we needed to follow a certain decoded rule to turn them into accurate numbers…. This methodology was really complicated but we made to convene a big group safely…. We really appreciated this (Y. Chen 2009a, 338).[16]

Graduates of Kansas State become known among islander engineers as "virus carriers" because they spread ideas of Taiwan identity and independence when they moved to other places across the United States (M. Chen 1992, 103–104; Chang and Tseng 2009, 199). One graduate in chemical engineering who became vice-chairman of United Formosans joined with his wife to help translate a portion of George Kerr's *Formosa Betrayed*, a 1965 indictment of the Chinese occupation that the KMT government had banned (Chang and Tseng 2009, 195–196, 244).

Yang Chung-Chang later reflected on the seeming correlation among islanders between engineering careers and independence activism. He observed that, for some, the attraction to graduate school in engineering was sometimes not prior to independence activism but actually served as one of its strategies. Islander students were finding, he said, that "becoming engineers would make it easier to get a job in the U.S. than through other professions" (Chang and Tseng 2009, 50; Y. Chen 2009a, 211). Such would make it possible to stay in the country after a student visa had expired. In contrast, many of those who came to seek higher degrees in political science, law, or other social sciences realized they would have difficulty finding positions in the United States, especially as university professors, if they carried a membership in an association advocating independence for

[15] Yu John Pang, interviewed by Chang Kuo-Hui, April 14, 2018.

[16] In addition to Yang, the major participants in the association were either engineering students or professors such as Wang J.-C. (王人紀), Chen S. K. (陳希寬), Huang Kim Lai (黃金來), Wang C. S. (王秋森), Strong Chuang (莊秋雄), Wang Neng-Hsiang (王能祥) and Hsu F. Y. (許富淵) etc.

Taiwan.[17] U.S. government policy on Taiwan was evolving but never included a risky endorsement of independence.

Yang also asserted that the commitment to discipline demanded by engineering training enabled United Formosans to maintain behavioral rules that prevented the organization from slipping into autocracy. United Formosans had "clear rules that included how its chairmen were to exercise their power and how long their terms were." Since "most of [United Formosans'] members had engineering backgrounds, it was natural that the rules were well-covered." Yang explained that members of a parallel organization in Japan "often teased us engineers" about the rules, 'but in reality our organization was run better than theirs.' When United Formosans later expanded into a worldwide organization, it adopted the bylaws of the U.S. chapter (Chang and Tseng 2009, 48–49; Y. Chen 2009a, 211).

By 1980, the rapid industrial emergence of Taiwan coupled with the waning of the KMT dream to retake the mainland helped justify a search for alternatives to the name Formosa. That year Chi Gou-Chung became a founding member of a new association of academic professors in the United States that would explicitly recognize islanders, including islander engineers. He reported that "we had a strong discussion at that time" about the name. It was no longer about Formosa. "[A]t that time," he said, "we were all Taiwan, come from Taiwan. We almost didn't say the word Formosa."[18] Convinced that it was time to use their own language and distinguish themselves from the West, the organizers made a slight but important shift in names. Rather than the noun "Taiwan" they decided to use the adjective "Taiwanese": North American Taiwanese Professors Association (Professors Association) (Liao and Chu 2005, 249).

Chi recounted how the leader of the new Professors Association, the University of Chicago biochemist Liao Shu-Tsung (廖述宗, 1931–2014), was questioned directly by KMT officials about the move. Liao had asserted that the adjective was appropriate because "for those who would like to approve Taiwan, make contribution to Taiwan," they "are Taiwanese." Ignoring the KMT attachment to China, according to Chi, Liao asserted that "Chinese" was "a vague term" that did not point specifically to the association's members, all of whom were from Taiwan.[19]

Then why not use "Taiwan"?, asked another KMT official. It was because the name "must have 'people.'" The new organization was an association of members who were explicitly committed to Taiwan as an entity separate from China, even though those members were in the United States, rather than to making Taiwan a means to China's future.

[17] The KMT government used a line on ROC passports to indicate that the passport holder was permitted to return, so the line worked as a kind of visa. If the passport expired while the holder was in the U.S., however, the ROC embassy would reissue one without that line if the holder's name appeared on the so-called blacklist. Then, like foreigners, they would need to apply for a visa to return to Taiwan.

[18] Chi Gou-Chung, interviewed by Chang Kuo-Hui, April 10, 2018.

[19] Chi Gou-Chung, interviewed by Chang Kuo-Hui, April 10, 2018.

The name referred to their attachment, committed to the land of Taiwan and its people, rather than to the land alone (Harrison 2006, 58). The organizers said that "of course people from other provinces [of China] could join," but with the familiar catch now in reverse. In this case, you had to "assert that you are Taiwanese."[20] Four years later, islander physicians in the United States followed suit and established the North American Taiwanese Medical Association.

As we indicated at the beginning of this chapter, the "T" word gained energy and legitimacy in naming when Lee Teng-Hui (1923–2020) became President of the Republic of China in 1988, following the lifting of martial law the previous year. Since Lee was born in Taiwan and served both as President of the Republic and as Chairman of the Kuomintang Party, to assert that one was Taiwanese, i.e., "of Taiwan," was no longer directly a threat to the Republic of China. It remained possible that it referred to a provincial identity, meaning "of Taiwan the province." President Lee's Ministry of Interior Affairs formally authorized new associations to use either "C" or "T" to represent the country in their names.[21]

For the first time, the term "Taiwanese" could refer, at least in principle, not only to islanders but also to mainlanders from Taiwan. In the United States, the new accessibility of the "T" word made it possible for islander engineers to begin presenting themselves simply and recognizably as Taiwanese engineers.

The Low-Profile Camp

In 1989, a year after Lee Teng-Hui became president, Chi Gou-Chung and a small group of fellow islander engineers working at Bell Labs rented three cars to go demonstrate in front of the United Nations headquarters in New York City. The student protests in Tiananmen Square in Beijing had just been violently repressed by the military, and the organizers wanted the United Nations and member countries to condemn and sanction the government of China. They expected that it would be easy to persuade many co-workers to join them since even the mainlander engineers and scientists had to be critical of brutality inflicted by Communist China. They calculated that roughly half of the 800–1,000 speakers of Chinese at Bell Labs were from Taiwan. However, organizing turned out to be enormously difficult, and they only recruited a dozen or so. All were islanders.[22]

Significant majorities of both mainlander and islander engineers in the United States did not join any of the voluntary professional associations, instead choosing a low profile. It can be difficult to account persuasively for a negative, an absence of action, especially absences in the past that involved distributed collections of people. At the same time, at

[20] Chi Gou-Chung, interviewed by Chang Kuo-Hui, April 10, 2018.

[21] In Chi's interview, he described how he benefitted from this change in the late 1990s while applying to create four new associations.

[22] Chi Gou-Chung, interviewed by Chang Kuo-Hui, April 10, 2018.

least some of the issues that were at stake for populations of islander and mainlander engineers, both separately and together, appear clear.

Surely some islander engineers, and probably all islander engineers sometimes, worried that KMT authorities were watching them. They feared that the public act of joining an islander association, which was prohibited in the Republic of China, could undermine their future return to Taiwan or pose risks to family members who stayed behind.[23] Their names could show up on a blacklist of expatriates who would not be permitted to return. As Chi put it in 2018, "Under the shadow of dictatorship, … we had to be very careful. We all had a police-general [a police agency in the KMT government] in our hearts. I couldn't talk about it casually…. It wasn't good for me."[24]

Indeed, for early islander engineers, not joining a Chinese association was itself sometimes a decisive action. Yeh Chiunn-Shyong (Eugene) (葉俊雄, 1951–), for example, had completed his bachelor's degree in electrical engineering in 1974 at National Cheng Kung University before he went abroad to the University of Southern California for his Ph.D. After finishing the doctorate, he worked a couple of jobs in software design before joining IBM in San José, California. Both at school and while working, Yen found that prior to Lee Teng-Hui's presidency most islander engineers were unwilling to assert a Taiwanese identity in public settings. However, by declining to join the Chinese Institute, "[t]hey passively resisted" the KMT government.[25]

On the other side, for many early mainlander engineers whose families had fled to the rural island province, that province was neither an ancestral home nor a promising locus for career development, and any return to the mainland seemed unlikely at best. Advanced education and work in the United States offered a chance for a new start, both professionally and personally. What benefit would there be in explicitly distinguishing oneself in the United States as a Republic-of-China engineer when the company one worked for might find itself in business relations with the People's Republic on the mainland? At Bell Labs, employees from different countries regularly sat together at lunch, e.g., a visible group from Japan. The dozen islander engineers who protested at the United Nations actually sat at a Taiwan table: they wanted to stand out as Taiwanese. There was, however, no such thing as a (Free) China table. A likely majority of mainlanders prioritized professional identities over Chinese identities in the workplace, finding it preferable to blend into a professional setting dominated by American colleagues.

It is certainly the case that many islander engineers also prioritized career development and professional identities, notably seeking job stability and income for themselves and their families that were much less available at home. For example, Chiu Chun-Pang (邱俊邦,1941–) came to the U.S. in 1967 to pursue a master's degree at the Oregon State University after completing undergraduate work in electrical engineering at Waseda

[23] Huang Kuan-Tsae and Roger Liao, interviewed by Chang Kuo-Hui, March 31, 2018 and April 3, 2018.

[24] Chi Gou-Chung, interviewed by Chang Kuo-Hui, April 10, 2018.

[25] Yeh Chiunn-Shyong (Eugene), interviewed by Chang Kuo-Hui, April 14, 2018.

(早稲田) University in Japan. He went on to found a semiconductor IDM (integrated design manufacturer) company in Silicon Valley, covering every step in the production of semiconductors from design through manufacturing to sales. He reported that while some islander engineers came to the U.S. in response to KMT oppression, "many more of them were driven by a desire to pursue a better professional career and a higher quality of life…They expected to create and implement innovative ideas."[26] They deliberately stayed away from issues in Taiwan's politics that defined opposing groups and wished to have a chance to demonstrate their technical talents.

Those islander engineers who subordinated their Taiwanese identities to their professional identities were actually producing a parallel or overlap with the islander engineers and engineering technicians in Taiwan who were animating a second Taiwan simply by doing their technical work. Those engineers and engineering technicians (see Chap. 4) helped their families initially by first securing well-paying positions with international companies and new mainlander companies in the electronics industry and then, for many, by building a multitude of new small companies in the many technical niches that appeared. The point is that neither the islander engineers in Taiwan nor those in the United States had to advocate publicly on behalf of Taiwan or proclaim their Taiwanese identities in order to resist continued subordination by mainlander engineers or their being identified as Chinese. They simply had to go to work, develop careers that carried upward income mobility, and maintain family commitments and responsibilities back home as best they could. Many clearly chose such a path.

We must also acknowledge a final potential pattern at work here that involved both mainlander and islander engineers and aligned them with working engineers trained in the United States and other countries. As we indicated in Chap. 1, educational programs in engineering tend to demarcate engineering work as technical, mathematical work that is distinguishable from politics, among other things. Doing engineering does carry a particular kind of political commitment in the sense that it promises to help everyone, but engineering curricula tend not to see this promise as politics. For both islanders and mainlanders trained in engineering, doing politics carried a specific implication that distinguished it from core engineering work—it meant taking positions in partisan conflicts. From this point of view, joining either a C association or an F or T association should be avoided because it constituted political action. It meant taking a position in a conflict. Many islander and mainlander engineers stayed away during the formative period described in this episode.

At the same time, do note that the C and F/T associations did not have politically partisan identities in the United States. They could accurately claim to be professional associations that simply co-existed with one another. Membership was not necessarily exclusive to one side or the other, and the two sides did not confront one another in any explicit way. To be sure, they competed for members and support for their visions. And, in

[26] Chiu Chun-Pang and Roger Liao, interviewed by Chang Kuo-Hui, April 13, 2018 and April 3, 2018.

principle, by joining any of them, one was working to serve or benefit something greater than oneself, in line with its particular commitment to omnibenevolence. One could join and not feel like one was betraying one's professional commitments. It is just that the visions of serving Taiwan that were enacted and embodied respectively by C and F/T associations contrasted sharply. Fulfilling the vision of one camp did, in fact, exclude the other's.[27]

Shifting Competition as Taiwan Becomes a Thing

In 1991 a group of about 80 islander engineers finally took the step of establishing the professional organization for engineers that included the name Taiwanese: North American Taiwanese Engineers' Association (hereafter NATEA). As we mentioned at the beginning of the chapter, the first president of NATEA, Yu John Pang, was a bachelor's and master's graduate in electrical engineering from Kansas State. He had come to the United States after completing an associate's degree at Taipei Institute.

Much had been happening in the electronics industry in Taiwan during the late 1980s. In 1987, the KMT government had taken the huge step of supporting the founding of a new private company to produce semiconductors for all sectors of the industry—Taiwan Semiconductor Manufacturing Company (TSMC). The government had previously attracted to Taiwan the mainlander electrical engineer Morris Chang (1931–), who had been raised in Hong Kong, earned degrees at MIT, and built a successful career at Texas Instruments in the United States.[28] Chang came to Taiwan initially to lead ITRI (Information Technology Research Institute), the government research organization that had remained largely in control by mainlanders during the rise of the electronics industry. Government support for TSMC signaled acknowledgement that all sectors of the electronics industry needed stable supplies of semiconductors, including the multitude of small and growing companies established by islanders (Fig. 5.1).

Given the small size of NATEA at the outset, an early goal was simply to exist and grow, drawing inspiration from islander families back home in Taiwan who were filling niches and contributing to the burgeoning electronics industry. As Yu John Pang put it, the main focus initially was "to keep improving in the spirit of small and medium-sized enterprises."[29]

Yu was well aware of the continuing presence in the U.S. of "C" associations, including a new one that we will turn to shortly. Accordingly, NATEA took care to ensure that the cost of participation in the association would not be a barrier to participation. Initially,

[27] Roger Liao and Huang Kuan-Tsae, interviewed by Chang Kuo-Hui, April 3, 2018 and March 31, 2018.

[28] While working at Texas Instruments, Chang also earned a Ph.D. in electrical engineering from Stanford University.

[29] Yu John Pang, interviewed by Chang Kuo-Hui, April 14, 2018.

Fig. 5.1 The ROC and Philips contract to build TSMC, 1986 (Front left second—Li Kwoh-Ting; front right fourth—Morris Chang) (Courtesy of The Central News Agency, Taiwan)

a yearly membership cost USD $50 and a lifetime membership USD $500, both low by engineering standards. Yet over time, the association concluded that even those fees posed something of a barrier. Members went to work finding sponsorships from private companies, and a single contribution of USD $50 became a membership for life.

By calling itself an association of Taiwanese engineers, NATEA did not explicitly demarcate islanders from mainlanders. It also did not define who counts as an engineer. "Anyone [could] come," said Yu, "as long as they identif[ied] with Taiwan." You were welcome, he elaborated, "no matter your politics or your background." It was just that identifying with Taiwan presupposed that members were "from Taiwan." By 1991, Taiwan itself was clearly a thing internationally, and even mainlanders within Taiwan were increasingly identifiable as "mainland Taiwanese" (Corcuff 2002). What happened, in fact, is that members of NATEA tended to be islander engineers who identified increasingly as "native" Taiwanese or, when in Taiwan, as "local" Taiwanese. The distinction between mainlander and native/local still carried a difference. In Yu's framing, "[T]here are two types of citizens within one country in Taiwan." Because of that difference, he continued, he "received lots of criticism from Chinese news media about founding NATEA, which bears with Taiwanese instead of Chinese."[30]

NATEA was founded in Silicon Valley, a strong indicator of the importance of electronics to the identities of many Taiwanese engineers (Saxenian 2006, 63). In order to prove itself attractive to engineers and attract them as members, NATEA organized regular seminars about different areas of the electronics industry. "At that time," said Yu, "the high-tech industry was not well-developed." It was only small and it "was very basic technology," so meetings took place only once every two or three months.[31] Yet the seminar proved to make crucial contributions to many Taiwanese engineers trying to plot their ways through a rapidly expanding and changing industry.

During the 1990s, the Association did indeed survive and grow. It added a seminar in biotechnology to its information sessions, and began to have contact in Taiwan with the National Science and Technology Council. That relationship would lead to a unique collaboration at the end of the decade.

Meanwhile, mainlander engineers in the United States adjusted in a wholly different way to the emergence of Taiwan as a thing via the electronics industry. One adjustment indicated the difficulty of the challenge: how to give the "credit" for Taiwan's increasing international visibility to the KMT government's concerted focus on the island itself without severing the island's attachments to the mainland? Accepting the label "Taiwanese" for themselves and their associations could risk disappearing the mainland origins of the mainlander government.

One response was to identify not with Taiwan as a bounded whole but with the tallest mountain in Taiwan—Yushan (玉山). Yushan is the fourth tallest mountain on an island in the world. Its peak of 3,953 m (13,966 feet) puts Taiwan behind only new Guinea, Hawaii, and Borneo. The adjusted vision was of an association based in Silicon Valley that would claim to represent engineers of Chinese descent and be led once again by mainland engineers, literally at its peak. Yushan is known as Monte Jade, purportedly because the snow at the top makes it shine like jade. The Monte Jade Science and Technology Association (Monte Jade Association) took shape in 1990, a year before the formation of the North American Taiwanese Engineers' Association.

Putting the mountain at its center enabled the Monte Jade Association to avoid prioritizing either Chinese or Taiwanese among its members. It was certainly "of Taiwan" but not limited "to Taiwan." The privileged status of mainland Taiwanese engineers over native Taiwanese engineers was preserved and reasserted, but no longer explicitly. That privilege blurred into a broader vision of a competition with Communist China to lead the world of Chinese descendants who worked in industry, especially electronics industries, in many countries. Monte Jade extrapolated the KMT vision to global terrain.

The rise of electronics industries in Taiwan during the 1970s and 1980s had made private industry a key site for negotiating the identity of Taiwan, both at home and abroad.

[30] Yu John Pang, interviewed by Chang Kuo-Hui, April 14, 2018.

[31] Yu John Pang, interviewed by Chang Kuo-Hui, April 14, 2018.

The rapidly-growing networks of relationships among start-up companies and technical talents in Taiwan and Silicon Valley attested to the prominence of this new marker of Taiwan's status and position. Monte Jade's full name affirmed the emergent symbolic significance of private companies by calling itself not an association of individual talents but a science and technology association, highlighting the membership of companies. It aimed to reach out and include experts who could help further private industry in Taiwan—where "private industry" meant initiatives linked to the large companies founded or supported by the KMT government in Taiwan. According to one islander engineer, Monte Jade emerged as "an association of big bosses."[32] Not only did the KMT government provide financial support for the association at the outset, but it also provided physical space. Monte Jade occupied the building in San Francisco that also housed the Republic of China's diplomatic mission in the United States.

The full name Monte Jade Science and Technology Association thus adjusted to the rise of Taiwan as a thing in the world by asserting that the confrontation between the Republic of China and the People's Republic of China was becoming an economic competition intensifying in private sectors—a competition that could be fought on multiple fronts and that the Republic of China could potentially win. Such could not happen, however, if the rise of industry in Taiwan was no longer clearly led by mainlander engineers that represented the Republic of China and its mainland heritage. Monte Jade dropped the "C" word but the competition continued in the United States between associations of mainland Taiwanese engineers and native Taiwanese engineers.

References

Andrade, Tonio. 2008. *How Taiwan Became Chinese: Dutch, Spanish, and Han Colonization in the Seventeenth Century*. New York: Columbia University Press.

Bucciarelli, Louis L. 1994. *Designing Engineers*. Cambridge, MA: MIT Press.

Chang, Yen-hsien 張炎憲, and Chiu-Mei Tseng 曾秋美. 2009. *Yi men liu mei xue sheng de jian guo gu shi 一門留美學生的建國故事 (A Story of the Nation's Founding by a Group of Students Studying in the United States)*. Taipei: Wu's Book Co., Ltd (吳氏圖書).

Chao, Che-sheng 曹哲生. 2007. "Hua fu guo jian hui li shi 華府國建會歷史 (CAPA History)." CAPA, https://sites.google.com/view/capadc/AboutUs/CAPAHistory. Accessed on Oct. 17, 2023.

Chen, Ming-Cheng 陳銘城. 1992. *Hai wai tai du yun dong si shi nian 海外台獨運動四十年 (Forty Years of Overseas Independence Movement)*. Taipei: Independence Evening Post (自立晚報).

Chen, Yi-shen 陳儀深. 2009a. *Hai wai Tai du yun dong xiang guan ren wu kou shu shi 海外台獨運動相關人物口述史 (An Oral History of People Related to Overseas Taiwan Independent Movement, Vol. 1)*. Taipei: Institute of Modern History, Academia Sinica.

Chen, Yi-shen 陳儀深. 2009b. *Hai wai Tai du yun dong xiang guan ren wu kou shu shi xu pian 海外台獨運動相關人物口述史續篇 (An Oral History of People Related to Overseas Taiwan Independent Movement, Vol. 2)*. Taipei: Institute of Modern History, Academia Sinica.

[32] Roger Liao, interviewed by Chang Kuo-Hui, April 3, 2018.

Corcuff, Stéphane. 2002. "Taiwan's "Mainlanders," New Taiwanese?" In *Memories of the Future: National Identity Issues and the Search for a New Taiwan*, edited by Stéphane Corcuff, 163-195. Armonk, NY: M. E. Sharpe.

Harrison, Mark. 2006. *Legitimacy, Meaning and Knowledge in the Making of Taiwanese Identity*. New York: Palgrave Macmillan.

Liao, Shutsung 廖述宗, and Chih-Pin Chuu 褚志斌. 2005. "Bei mei zhou tai wan ren jiao shou xie hui (NAPTA) chuang hui shi zhou nian ji shi 北美洲台灣人教授協會(NATPA)創會十周年紀事 (10-Year Anniversary of NATPA)." In *Zi jue yu ren tong: 1950–1990 nian hai wai tai wan ren yun dong zhuan ji* 自覺與認同*: 1950–1990*年海外臺灣人運動專輯 *(Self-consciousness and Political Identity: Overseas Taiwanese for Taiwan Independence from 1950 to 1990)*, edited by Yen-hsien Chang 張炎憲, Chiu-Mei Tseng 曾秋美 and Chao-Hai Chen 陳朝海, 247–280. New Taipei: Wu's Book Co., Ltd. (吳氏圖書).

Saxenian, AnnaLee. 2006. *The New Argonauts: Regional Advantage in a Global Economy*. Cambridge, MA: Harvard University Press.

Su, Kuo-Hsien 蘇國賢, and Wei-Hsin Yu 喻維欣. 2007. "Tai wan zu qun bu ping deng de zai tan tao: Jie shi ben sheng wai sheng zu qun cha yi de suo jian 台灣族群不平等的再探討: 解釋本省/外省族群差異的縮減 (When Social Reproduction Fails: Explaining the Decreasing Ethnic Gap in Taiwan)." *Tai wan she hui xue* 台灣社會學 *(Taiwanese Journal of Sociology)* 39 (December): 1–63.

6 The Separation Crosses the Strait

During the 1990s, the distinction between mainlander engineers and local engineers in Taiwan appeared to blur. The blurring was especially salient for engineers involved in the highly-distributed industries of electronics and, increasingly, information technology. A key indicator was the enthusiastic participation of both in extending Taiwan-owned technologies and companies across the Taiwan Strait to mainland China.

The apparent decline in sharpness took place within rapidly expanding attention across Taiwan to what became known as "cross-strait relations." On the mainland side, the Chinese government had recently begun permitting provinces and municipalities to invest in privately-owned industries (Leng 2005, 68–69). It also authorized direct travel to Taiwan for the first time since the end of the civil war (Sutter 2002, 523–524). On the Taiwan side, both large firms supported by the KMT government and led by mainlanders and small firms owned and operated for the most part by local families rushed across the Strait to set up production facilities for a range of technologies and products.

All the companies that traveled were pursuing at least two sets of opportunities. One was lower labor costs, for over the previous two decades wage rates in Taiwan had grown to nearly ten times those on the mainland (Bishop and Chiou 2004, 554).[1] Lower labor costs did not necessarily mean lower skills. Taiwan-based companies that designed integrated circuits, for example, were looking for engineering talents in China because the local supply could not keep up with the increasing demand.[2] Another was the likelihood

[1] For example, the owner of a molding shop located in Hsinchu amidst a dense maze of such shops, explained how his family had set up and was managing a related shop across the Strait for just this reason—labor costs a tenth of what he was paying at home. After stipulating anonymity, he peeled a piece of paper off of one machine to reveal that it was producing plastic parts for Apple. Anonymous, interviewed by Gary Downey, March 5, 2003.

[2] Wu Chin-Ju, interviewed by Chang Kuo-Hui, November 20, 2023.

K.-H. Chang et al., *Engineers and the Two Taiwans*, Synthesis Lectures on Global Engineering, https://doi.org/10.1007/978-3-031-59766-4_6

that extending manufacturing operations to the mainland would enable Taiwanese companies to gain direct access to suppliers and customers in China, giving them footholds on the mainland (Sutter 2002; Leng 2005).

At the same time, the actual work of transporting the production of electronics and information technologies to the Chinese mainland carried different meanings and implications for the careers and lives of engineers in Taiwan with contrasting geographical identities—mainlanders and locals. By establishing and maintaining a physical presence on the mainland, both were also enacting images of and commitments to Taiwan in relation to China, and established differences between the two main versions of that relationship would not dissolve away. Indeed, those differences would actually gain additional contents and enhanced significance for engineers over the next three decades. And engineers from the mainland would enact yet another image of China and Taiwan as the Chinese government increasingly asserted its own vision of the meanings and implications of cross-strait relations.

The Multiplication of Local Engineers

During the 1990s and early 2000s, enrollments in engineering degree programs at Taiwanese universities basically exploded. The total number of students enrolled in bachelor's programs increased from approximately 39,000 in 1990 to nearly 226,000 in 2005. The number of master's degree-level students also increased significantly, from around 5,600 in 1990 to more than 40,600 in 2005. At the Ph.D. level, while the absolute number of students who chose to go abroad did continue to increase, greater percentages of prospective students were turning to pursue this degree at local universities, totaling from 1,700 in 1990 to 11,400 in 2005 (Department of Statistics 2022).

Although enrollment data in Taiwan's universities were not coded to distinguish mainlanders from locals, a significant majority of the increase in engineering students were likely locals.[3] By the 1990s, most mainlander students were second-generation, meaning they were born in Taiwan. Among the total population of college-age men and women (ages 15–24) during this period, the number of locals was six times the number of second-generation mainlanders (Census Office of the Executive Yuan 1992, 622–649). Also, prior to the 1990s, local students interested in engineering had found themselves mostly limited

[3] A rare exception is rosters of Ph.D. degrees awarded prior to 1984 that listed the geographical identities of recipients, self-identified. Between 1970 and 1973, a total of three islanders and six mainlanders received the Ph.D. in engineering, whereas during 1974–1983, engineering Ph.D. degrees were awarded to 49 islanders and 21 mainlanders Education Yearbook Compilation Committee of the Ministry of Education (1985, 1976). Early on, the KMT government did compare the numbers of local students overall in higher education with Japanese students in 1944 and mainlander students in 1949. The number of Japanese students in 1944 was four times that of local students: 360 to 85. In 1949, when mainlanders were arriving en masse, the number of local students was double the number of mainlanders: 803 to 383. See Department of Education (1950).

to pathways through junior college curricula and into support-level technical positions. During the 1990s and early 2000s, a dramatic expansion took place in opportunities to look beyond junior colleges and purse higher education in engineering.

The 1990s are known across Taiwan as a period of democratization and indigenization, starting with the loosening of restrictions after the end of martial law in 1987 and marked especially by the first direct presidential election in 1996. Demands by Taiwanese locals for rights, recognition, and economic opportunities appeared in a multitude of social and political movements and official initiatives. These movements and initiatives sought to rectify inequities across a range of concerns, from environmental protection, consumer protection, women's rights, and indigenous rights to calling attention to the '228' abuses, support for victims of the White Terror, redesign of electoral procedures and representation, and expanded access to education (Keating 2008). Their outcomes added new dimensions to what it meant for Taiwan to be a thing in the world.[4]

Prospective local students interested in engineering benefitted especially from a redesign of education policies by the KMT government to make higher education more accessible. Following a two-year study, the government began in 1994 to invest heavily in junior colleges in order to warrant elevating them to university status, and their graduates to university graduates (Rau 1998; C. Wu and Chien 2014). The effort largely worked. By 2005, fully 68 of 162 junior colleges had been upgraded to universities, with approximately 40 including engineering departments, and as access to bachelor's degrees in engineering expanded, students followed the upward flow (Department of Statistics 2006a, 451–452; 2004; D. Lin 2006, 94). Also, with fewer junior colleges, enrollments in them dropped accordingly. Between 1990 and 2005, the number of students enrolled in junior colleges decreased by nearly 75%, from around 154,000 to just under 41,000 (Department of Statistics 2022).

Not surprisingly, as engineering students responded to the new opportunities, their curricular interests prioritized electrical/electronics/information engineering and computer science. Students were clearly seeking to participate in the explosion of industries that had made Taiwan a thing in the world and was then contributing to promising expansions across the Taiwan Strait and into China (Smart and Hsu 2004; Fuller 2008).[5] By 2005, they comprised more than a third of enrollments in engineering disciplines (Department of Statistics 2006b, 119–120). Education in civil engineering still prepared graduates primarily for work on public infrastructures and construction in the private sector, and chemical engineering pointed students mainly toward chemical plants. Neither arena of engineering work was traveling to the mainland at the level of workplaces for electrical/

[4] All of these movements had researchable implications for engineers and engineering in Taiwan. Our focus here remains on the trajectories of dominant images and practices of engineering and the geographical identities of engineers.

[5] It has been reported that, by the late 1990s and early 2000s, investments in China's electronics industry had grown to share roughly half of the total amount of the investments across the Strait from Taiwan to China (Fuller 2008).

electronics engineering and computer science. Mechanical engineering remained attractive to increasing numbers of students, for capabilities in machine design and maintenance made mechanical engineers integral participants in virtually all forms of industrial production, including electronics and information technology. Still, although enrollments in mechanical engineering increased substantially during the 1990s and early 2000s, the clear preponderance of student interests aimed toward electronics and computer science.

The Return of Engineers

Alongside the rising status and increasing numbers of engineers trained in Taiwan, a second key development for local engineers during the 1990s and early 2000s was the opportunity for those working abroad to return home. The termination of martial law and the government's blacklist of dissidents offered persuasive evidence that local engineers could embrace work for Taiwan without fear for themselves or their families. The government of Lee Teng-Hui emphasized the point by actively inviting prominent engineers to return, including those who counted themselves as locals.

Chi Guo-Chung (紀國鐘, 1946–) was one such engineer. In 1990, he accepted a government invitation to bring back to Taiwan the knowledge and experience he had gained from years working at Bell Labs. Chi took some time to decide whether or not to return, informed by friends and colleagues who were wary about the possibility of future war with China. The father-in-law of an islander friend at Bell Labs, for example, actually warned, "Don't go back." He was himself not an islander but a mainlander, a former KMT general. His daughter had married Chi's islander friend and brought her parents to live with them. The two parents had personally experienced the horrors of the civil war in China. Chi reported how, in great detail, they "told me … how cruel the Communist Party was."[6]

In an extended interview, Chi explained his decision to return by highlighting a desire to fulfill family responsibilities. He "came back to Taiwan to be the master of [his extended] family." If war with China did in fact happen, he added, he "would send his wife and children back to the United States." And to ensure they would be able to live there without him, he kept his green card, marking him as a permanent U.S. resident. Chi also "didn't move the money" from the United States to Taiwan.[7]

For Chi, recognizing and accepting the pull of family responsibility turned out to also indicate a more pervasive connection to the land and people of Taiwan that was difficult to maintain in the United States. After returning, he found himself regularly insisting that he was not an outsider. Chi remembered saying to mainlanders and locals alike, "I don't need you to greet me." Rather, because of his long family history on the island, he saw

[6] Chi Gou-Chung, interviewed by Chang Kuo-Hui, April 10, 2018.

[7] Chi Gou-Chung, interviewed by Chang Kuo-Hui, April 10, 2018.

himself in an important sense as already there. "I'm a local," he continued, "[who] just went out and wandered for a while and then came back."

Chi was interested in involving himself in what he called "public affairs," which meant gaining a formal position in government. He accepted a post at ITRI but left after a few years to take a teaching position. His colleagues at ITRI had told him that his career was compromised because he explicitly supported the Democratic Progressive Party (DPP). In the terms used in this book, he had not only exceeded his proper engineering identity by embracing partisan politics but also made his geographical identity the ground for a partisan challenge to the ruling Kuomintang Party.[8] He was a local engineer contributing to a Taiwan that was/should be separate from China.

Another early returnee, Wu Chin-Ju (吳欽智, 1946–) came back from the United States to work in the private sector, managing a new subsidiary of Acer formed to design integrated circuits. After obtaining a Ph.D. in material science at the University of Southern California with a focus on semiconductor technologies, he had initially taken a job at Rockwell in that area. With a growing interest in design, he had moved to Quasel, an IC design company. There, he said, "Although I had to work 100 h a week, I was engineering satisfied." When Acer's founder Stan Shih invited him to return in 1987, he jumped at the chance, for two reasons. "I knew I was less satisfied with working for big companies," he said, "[and] was looking for an opportunity to run a business." Running a subsidiary of Acer was "an industrial opportunity for me as an IC design engineer," in part because Taiwan was building foundries for contract chip-making. At the same time, "I also felt a strong attachment to Taiwan. It was not just because my original family was there but I often felt alienated from the United States."[9]

A third early returnee, Lin Mou-Shiung (林茂雄, 1952–), returned in 1988 to help Taiwan Semiconductor (TSMC) get started manufacturing semiconductors. After completing a B.S. in physics at National Taiwan as well as his required military service, Lin had left for Harvard University in 1977 to pursue his Ph.D. in applied physics. After working initially for IBM, he had transferred to Bell Labs to study semiconductor technology, while also hoping for opportunities to help produce them in Taiwan. The offer from TSMC more than fit the bill: "I just could not wait to work with them." Once again, there was also a strong attachment to Taiwan. Although Lin stayed away from partisan politics, he did take meals with Chi Gou-Chung at the Taiwan table. And knowing that President Chiang Ching-Kuo had recently passed away and that Lee Teng-Hui's government might now reach out to islanders, he anticipated that Taiwan could now "become a democratic country." Returning to Taiwan might help him realize a longstanding wish to devote himself to the well-being of locals. At Bell Labs, a mainlander colleague made the point explicit: "You should return since you are a Taiwanese, unlike me."[10]

[8] Chi Gou-Chung, interviewed by Chang Kuo-Hui, April 10, 2018.

[9] Wu Chin-Ju, interviewed by Chang Kuo-Hui, November 20, 2023.

[10] Lin Mou-Shiung, interviewed by Chang Kuo-Hui, November 22, 2023.

The early flow of returnee engineers to Taiwan became a reverse brain drain in the 1990s and early 2000s (Saxenian 2006, 147–153; M. Wu and Yang 2022, 25). During the period 1986–1996, more than 40,000 people returned to Taiwan from the U.S. By 1989, about 500 engineers had returned to work in the Hsinchu science park, with the total number reaching more than 4,000 by the early 2000s (Saxenian 2006, 148, 151; Watanabe 1989; Hsinchu Science Park Bureau 2006, 45).

Thousands of the returnees were local engineers who came back from the United States, Europe, Japan, and elsewhere, sensing unprecedented opportunities to work in Taiwan without restrictions on rank, position, or daily life. For engineers who had previously been denigrated as islanders and could now view themselves as local Taiwanese, returning to Taiwan enabled them to forge new intersections between career and family identities. And since Taiwan was by then recognized internationally as a site of industrial production, those who returned specifically to help build industries in electronics and information technology could also view themselves as helping Taiwan elevate its standing further, potentially into a world of countries as a member. Put inversely, in the 1990s it finally became possible for engineers overseas to overcome the sense that career interests and family responsibilities had to pull one in opposite directions.

At the same time, the emergence of Taiwan as a thing that hosted bustling networks of large and small companies also meant that returning local engineers were not necessarily welcomed back with open arms by other local engineers, business owners, and government officials. Their identities and priorities were subject to question. Who were these engineers who had wandered for a while and suddenly wanted to resume, and now presumably influence, life in Taiwan? Were they actually coming to help Taiwan in some way or only to help themselves and their families? Even if both, should they expect special acknowledgment or recognition? The electronics and information technology industries had evolved rapidly while they were abroad, meaning that every producer, small and large, had to find technical strategies and pathways to remain viable. Was the return of many experienced engineers adding anything more than competition that would make productive success by other locals more difficult to achieve and maintain? As the early returnee Chi put it, echoed by two later returnees including one who must remain anonymous, "My generation received better treatment" than those who followed later.[11]

One recurring challenge for local returnees who had been working in shareholder-owned corporations in the United States or elsewhere was that they had to figure out how to fit comfortably into local networks of family-owned businesses or whether they even wanted to do so. As late as 2018, as many as 80% of listed companies in Taiwan were still family businesses (Z. Lee 2018, 316). For returning local engineers, integrating their career and family paths could be a more demanding experience than they had anticipated or desired. Many were unprepared to give up authority based on technical competence and performance to authority based on family status, even as they contributed skills in produc-

[11] Chi Gou-Chung, Anonymous, and Wu Chin-Ju, interviewed by Chang Kuo-Hui, April 10, 2018, April 3, 2018, and November 20, 2023.

tion and management.[12] As observed by one senior local who had stayed in the United States, "some Taiwanese engineers … are dissatisfied with the local family-first business model and hope to see changes."[13]

A significant number of mainlander engineers also returned. For them, a Taiwan that had become a thing unto itself internationally made returning to it an uneven prospect. After all, their families had only been connected to Taiwan for little more than a generation. And for a long time, as Chi explained, "[t]he Kuomintang's mentality" was that if you studied abroad "you have drunk foreign water so we hold you high…."[14] For mainlander engineers who wanted to connect to ancestral homelands on the mainland, a Taiwan that was increasingly gaining an identity separate from China could make working in, and for, Taiwan even more uncertain than it had already been.

Still, for many mainlander engineers who had built careers in the U.S., founding and participating in Taiwan-based companies that marketed in China and established divisions in China could be both lucrative personally, something highly valued in the U.S., and promise new links to the mainland. Families would not have to move to Taiwan to make the connection happen, and those that did would not have to make the move permanent. Furthermore, participating specifically in the expanding flow of electronics and information technologies into China just might provide footholds that keep open the possibility of unification by attracting China towards Taiwan rather than the other way around. So while local engineers were able to feel a new pull from Taiwan itself, mainlander engineers could feel a pull from Taiwan-owned companies that extended their boundaries across the Strait to the mainland, especially those companies that had been created with support from the KMT government.

Crossing the Strait on Two Pathways

In 1992, the KMT government announced a new economic policy allowing, and encouraging, open investment in China. Electronics companies moved quickly to build manufacturing facilities on the mainland. They continued to take orders for products to export, but over the next decade those exports were shipped increasingly from the mainland. When the government began in 1999 to formally track the proportions of exports from Taiwan-owned firms that were actually manufactured outside the island, the initial calculation was slightly more than 12%. By 2016 that number would grow to more than 54%. Nearly two-thirds of all exports that year fell into the three categories that pertained most to electronics and information technology: Information, Communication, and

[12] Roger Liao, Yeh Chiunn-Shyong (Eugene), Huang Kuan-Tsae, Chiu Chun-Pang, and Rockwell Hsu, interviewed by Chang Kuo-Hui, March–April 2018. See also M. Lin (2023, 223) and Tung et al. (2016).

[13] Anonymous, interviewed by Chang Kuo-Hui on April 14, 2018.

[14] Chi Gou-Chung, interviewed by Chang Kuo-Hui, April 10, 2018.

Audio-visual products (30%); Parts of Electronic Products (27%); and Optical, Precision, Instruments (5%). And within those categories, 93%, 47%, and 47% of the totals were produced and exported outside of Taiwan, respectively, with the vast majority of outside production taking place in China (Department of Statistics 2017).[15]

These dramatic increases in quantities indicate that, during the 1990s and early 2000s, both large and small firms in the electronics and information technology industries extended operations to the mainland. While the mainlander and local engineers who participated found themselves adding operations on the mainland to existing operations in Taiwan, their simultaneous movements did not indicate some kind of reconciliation between the contrasting visions of Taiwan as a thing in the world. The engineers traveled across the Strait not together as a homogeneous group but along two distinct pathways, one maintaining centeredness and control in Taiwan and the other building productive power with mainland China. The manufacture of semiconductors became a key site of tension between them.

During the 1990s, large government-supported and mainlander-controlled companies in Taiwan developed a new industrial strategy to maintain their visibility and relevance internationally. Knowing that all production operations in electronics and information technology, small and large, needed access to computer chips, they decided to focus narrowly on providing foundries for the manufacture of semiconductors. A key stimulus was the realization that companies producing semiconductors in the U.S., such as Texas Instruments and Intel, had started diversifying their businesses and avoided limiting themselves to manufacturing semiconductor devices. They were concentrating increasingly on designing ICs and semiconductor devices for ever more challenging and complex uses.

TSMC reoriented its operations away from designing computer chips to, more simply, producing them under contract with design firms located elsewhere. In a sense, the company was replicating the 1960s–1970s domestic strategy of locating low-cost production in Taiwan. But this time it focused not on consumer goods for end users but on the key semiconductor manufacturing knowledge that provided for entire industries. Also, the reductions in cost in this case would be achieved less via low-wage workers and more via high volumes of production and a single-minded, highly specialized technical commitment to keep increasing the numbers of components and the computing capabilities that could be squeezed onto a single chip. The new industrial strategy would initially require high-level engineering expertise in research and development centering on manufacturing (C. Chang and Yu 2001; Lüthje 2006). And since semiconductor manufacturing requires large amounts of water and electricity, the shift involved taking on all the environmental issues that would come with it (S. Chang et al. 2006; Smith et al. 2006).

[15] By 2016, nearly 95% of Taiwan's exports that were from foreign orders but produced "overseas" were produced in China.

In the midst of the rush to the mainland, an open question was whether semiconductor foundries would go there, too. Would TSMC and other companies that were narrowing their focus to foundry-work gravitate toward mainland China, taking that productive power with them? Would the KMT government permit or encourage this? Would building foundries on the mainland enable the Chinese government to gain access to production knowledge of the computer chips that brought them to life? Who and what could go to China, or not?

The official practices involved in reaching out to China started to exhibit an explicit tension in 1996 through a change in government when Lee Teng-Hui became president. Shortly after his election, Lee spoke to a conference of Taiwan's industry and commerce associations, most of whose members were investing actively in China. The energy for crossing the Strait had been so great that by 1997 Taiwan's bilateral trade with China would constitute more than 11% of its total trade in the world, up from only 1% in 1981 (Mainland Affairs Council 1999). And from the point of view of the mainland, Taiwan was on its way to become the third largest outside investor in its domestic industries, behind only Hong Kong/Macau and Japan (Gao 2003).

President Lee, the first president to be born in Taiwan, disappointed attendees at the conference by counseling patience and wariness of impetuosity (T. Lee 1996). His new government raised a concern that flows of technology, knowledge, capital, and people might not extend Taiwan's productive power across the Strait but have the opposite effect, shifting productive power to industries on the mainland. A few months after the conference, Lee's administration began formalizing this concern by overtly discouraging Taiwan's companies from investing across the Strait in both high-tech industries, including electronics and information technology, and their supporting infrastructures. In particular, it stipulated that all new investment applications to China would have to be screened (Mainland Affairs Council 2002) (Fig. 6.1).

For its part, TSMC took dramatic steps to expand its domestic foundry capacity and maintain its centeredness in Taiwan by fully acquiring both Acer's semiconductor memory foundry and another competitor, Worldwide Semiconductor Manufacturing Corporation (Worldwide Semiconductor). Acquiring Acer's semiconductor capability had the effect of sharply separating contract chip-making in Taiwan from the work of both upstream chip design and downstream manufacturing of instruments and consumer products, and promised to maintain centeredness in Taiwan. Acquiring Worldwide Semiconductor, however, challenged and sharpened a tension with the second pathway. Its CEO was a mainlander engineer who was clearly committed to shifting the center of gravity in semiconductor production toward China.

Richard Ru-Gin Chang (張汝京, 1948–) had followed a common pattern of would-be engineers who had been born on the mainland. Only two years old when his parents fled to Taiwan in 1949, he went on to gain admission to National Taiwan University and complete a B.S. in mechanical engineering. He then sought advanced education and work in the United States, completing a Ph.D. in mechanical engineering at Southern Methodist

Fig. 6.1 Lee Teng-Hui (standing) at National Unification Committee, 1997, counseling patience and wariness of impetuosity (Courtesy of the Central News Agency, Taiwan)

University and finding a position at Texas Instruments. Over a period of two decades, he helped Texas Instruments build foundries for the production of microchips.

Chang later reported that during his time at Texas Instruments he had long dreamed of building a semiconductor industry for China (M. Hsu 2000, 177; Editorial Board 2002; Miller 2022, 177). The opportunity to realize that dream came in 1997 when he moved back to Taiwan to become the founding CEO of Worldwide Semiconductor. The geographical descriptor "worldwide" in the company's name fit the 1990s strategy among mainlander engineers that the Monte Jade Association articulated explicitly in its name. Among those standing at the top of Monte Jade, the company sought to position itself as a leader of the global diaspora of ethnic Chinese. However, the acquisition of Worldwide Semiconductor by TSMC only three years later came as a surprise to Chang and took place over his objections.

So Chang left Worldwide Semiconductor and crossed the Strait. Gathering financial support from Goldman Sachs and the Chinese government, he established a new foundry in Shanghai, the Semiconductor Manufacturing International Corporation (SMIC). In this case, use of the descriptor "international" in the company's name fit the Chinese government's increasing efforts to expand its influence beyond its boundaries through productive power in industry. Chang successfully brought approximately 300 foundry engineers with him from Taiwan. These included both former engineers from Worldwide Semiconductor and engineers from TSMC who wanted to work and build connections on the mainland (B. Chen 2008).[16]

Maintaining Productive Power in Taiwan

The first pathway for transporting the new technologies across the Strait, maintaining centeredness and control in Taiwan, gained a burst of energy in 2000 when the KMT lost control of the presidency to the Democratic Progressive Party (DPP). For the first time in the emergence and evolution of engineers and engineering in Taiwan, native/local Taiwanese engineers gained explicit formal opportunities to participate in governance. The year before the election, the DPP had approved a resolution asserting that "Taiwan is an independent and sovereign state." It envisioned formalizing a cross-Strait relationship with China that would be "reciprocal rather than discriminatory, peaceful rather than confrontational, and equal rather than subordinate" (Democratic Progressive Party 1999). To the lead party in the newly-elevated government, Taiwan should no longer be treated as a province of China and should operate in the world as a country alongside other countries.

The new government reached out for help to native Taiwanese professionals working in the United States, including many engineers. Yeh Chiunn-Shyong (Eugene) (葉俊雄, 1951–) was a case in point. "When the DPP took the president's office," he said, "I was reached by friends who reported on new opportunities to find a job in Taiwan." Yeh had been thinking of returning to Taiwan because of personal connections as "most of my original and siblings' families were in Taiwan." He initially took a leave of absence from IBM to work in the research Center for High-speed Computing at the National Science and Technology Council. He later resigned from IBM in order to accept the position of Deputy Center Director.[17]

The National Council took the lead in recruiting engineering researchers from the United States. It reorganized six research centers, merging them into a new national agency, the National Applied Research Laboratories (National Labs) (National Applied Research Laboratories 2023).[18] Members from the North American Taiwanese Engineers' Association (NATEA; here Engineers' Association) as well as from the North American Taiwanese Professors Association and a new industry association enthusiastically took advantage of opportunities to return to positions at the "national" level.

[16]The KMT government's original state-owned semiconductor manufacturer, United Microelectronics, also dropped its work in chip design and narrowed its focus to contract foundry work. However, it remained committed to expanding connections with the mainland. Within a few years of Lee's election, it was working to bypass his government's efforts to limit the shift of productive power in the new industries to China (Geppert 2005).

[17]Yeh Eugene Chun-Hsung, interviewed by Chang Kuo-Hui, April 14, 2018.

[18]Four additional centers were established later.

The National Council also reached out to the Engineers' Association to formalize flows of engineering knowledge to support Taiwan-centered companies, now including the small and medium-sized companies organized by locals. For the first time, the Council provided direct financial support to that association of predominantly native Taiwanese engineers, even while continuing to support the Chinese Institute of Engineers and the Chinese American Academic and Professional Society.[19] In 2003, it joined with the Engineers' Association to establish the U.S. Taiwan High-tech Forum (High-tech Forum), which called itself a "venue to bridge engineers and scientists on both sides of the Pacific Ocean on technological issues."[20] According to a key founder, Rockwell Hsu (徐名璋, 1963–), the High-tech Forum quickly became a kind of "playground" where Taiwan-centered engineers in the U.S. could meet people, build business-oriented networks, and find entrepreneurial opportunities (R. Hsu 2021).[21] In other words, it was a way for the DPP government to stimulate the further development of small and medium-sized start-up companies in, especially, electronics and information technology.

Although a native Taiwanese himself, Rockwell Hsu had some mixed feelings about the initiative. He first went to the U.S. in 1988 to complete both undergraduate and graduate degrees in electrical engineering at Florida Institute of Technology. He reported in an interview that at the point of graduation, "[I] began to identify myself Taiwanese." He had to apply for an H1 visa and export license from the U.S. Department of State to accept a position with the U.S. defense contractor Loral Corporation located in Silicon Valley. The immigration agent and export license agent thought that the Republic of China was mainland China rather than Taiwan, which would have denied him eligibility for a visa. Hsu reported that when he said, "I am from Taiwan," the agent replied, "Ok, you have a chance." Hsu had previously encountered and joined the Engineers' Association as an extension of his prior membership in IEEE, the U.S.-based professional society for electrical and electronics engineers. He chose to help the Engineers' Association establish a working relationship with the National Council. But in the context of the change of governance in Taiwan, he "worried whether [the Association] was a political organization promoting Taiwan independence."[22]

New energy from native Taiwanese engineers for start-ups in electronics and information technology based in Taiwan also came from informal networking and coordination between new industry associations in both Taiwan and the U.S. In 2001, Chi Guo-Chung worked with Huang Der-Ray (黃得瑞, ?–2019] in Taiwan to form the Taiwan Industrial Technology Association to support, especially, small and medium-sized enterprises. The organization then reached out to engineers in Silicon Valley to establish a parallel association. Two years later the Taiwanese American Industrial Technology Association—Silicon Valley (Industrial Technology Association) came into being.

---

[19] Huang Kuan-Tsae, interviewed by Chang Kuo-Hui, March 31, 2018.

[20] See also NATEA (2023).

[21] Rockwell Hsu, interviewed by Chang Kuo-Hui, April 22, 2018.

[22] Rockwell Hsu, interviewed by Chang Kuo-Hui, April 22, 2018.

Although led by engineers, the Industrial Technology Association sought any and all members who could contribute to founding and building new companies. It focused in particular on recruiting young Taiwanese professionals, not only to join but also to lead. Having an explicit focus on building new private companies would enable it to present itself as apolitical on the issue of independence for Taiwan. At the same time, the Industrial Technology Association could respond to and balance out the force of the Monte Jade Association, which was seeking to position large Taiwan-based companies as leaders of the Chinese diaspora worldwide.[23] Without dismissing or denigrating the initiatives of large companies, the Industrial Technology Association sought to dramatically enhance research and development efforts to support the burgeoning arena of small companies.

During these early years of the DPP administration, members of the KMT party complained that the DPP was actually recruiting independence activists. They called the rebuilt National Science and Technology Council the "drug den," cleverly trading on the fact that the Mandarin translations of "drug" and "independence" are pronounced the same. One KMT congresswoman produced a list that questioned the duties of advisers brought in to serve in the Science and Technology Policy Research and Information Centre for the National Labs. The then-minister of the National Labs asserted that the advisers at issue were all qualified to serve as technology policy experts, but also acknowledged that they did indeed do some work that was irrelevant to technology policy (Legislative Yuan 2006). For our purposes here, whether the imported research advisers were advocates of independence or not, it was clear that the new government focused on building productive power within Taiwan as a thing unto itself.

With so many Taiwan-based companies managing manufacturing facilities and seeking markets in China, policymakers for the DPP government looked for ways to keep relations across the Strait positive while maintaining control in Taiwan over the flows of engineering knowledge. After all, by 2010 roughly two percent of Taiwan's population of twenty million people were working on the mainland (DGBAS 2017). One justification that the DPP offered for maintaining relations was that Taiwan could become a role model for the Chinese people through its successes in industry, which now included small and medium-sized companies. Taiwan could demonstrate the benefits and possibilities of "free and open" investment in start-up companies and the social benefits of increased status for contributing to the world electronics and information technology goods that were in increasingly high demand.

In line with this justification, the DPP government loosened Lee Teng-Hui's cautionary approach to "overcome impetuosity and exercise patience (or be patient, go slow)," renaming it "proactive liberalization with effective management." Effective management could include further expansions in manufacturing and trading relationships. However, it could not include transfers of engineering knowledge sufficient to shift power and

[23]Yang, Chi-Hang, interview by Chang Kuo-Hui on June 29, 2018. In addition to the Industrial Technology Association, the Washington D.C.–based Science & Technology Association was founded in 2002.

control over electronics and information technology to China. After 2005, due to deteriorating relations with the Chinese government, the DPP reframed the policy again as "proactive management with effective liberalization," with an increased emphasis on management (Sam 2007; Karackattu 2008; Wong 2005).

Concerns about a shift in productive power continued to center on the manufacture of semiconductors, the engines of electronics and information technology. In 2002, the DPP government released an industry policy called the "Two Trillion and Twin Star Development Program (兩兆雙星)" that focused especially on semiconductors for computers and computer displays (National Development Council 2002). As part of this new policy, the Ministry of Economic Affairs would review and monitor projects by Taiwan-based companies that aimed to build semiconductor foundries in China.[24] It became known as the "N-minus-1" policy because it stipulated that manufacturing plants to be built in China would produce wafers that were at least one generation and an additional six months behind those produced in Taiwan (Ministry of Economic Affairs 2002a). In one example, once the foundries in Taiwan were able to produce 12-inch wafers for more than six months, they could begin manufacturing 8-inch wafers on the mainland. Control over cutting-edge computing power was supposed to remain in Taiwan.

For engineers and others traveling this first pathway across the Strait, the transfer of engineering knowledge from Taiwan-based companies to China-based companies posed a potential risk to the future identity of Taiwan. The former TSMC senior local engineer Lin Mou-Shiung and two close colleagues, for example, provided confidential technical advice to the government about how to defend a policy that would prevent Taiwan-based companies from building advanced semiconductor foundries in China. "Many years later," Lin reported, "every time [my colleague] and I recall the event, we consider our help in the battle [to achieve such prevention] was meaningful in our lives and we are proud of it" (M.-S. 林. Lin 2023, 218). If we let the local semiconductor industry to relocate to China, "we would lose Taiwan's economic competitiveness, technological advancement and young people's opportunities for decent jobs."[25]

In 2002, the DPP government stiffened the N-minus-1 policy by capping at three the number of Taiwan-based companies that could transfer to the mainland the knowledge to produce 8-inch wafers (Mainland Affairs Council 2003). It also initiated a novel project to support start-up companies and recruited Huang Kuan-Tsae (黃光彩, 1953–) to participate. As Huang put it, "I consulted for the DPP administration's project called ABCDE, for the enhancement of Taiwan's supply chain in the manufacturing industry" (Ministry of Economic Affairs 2002b). During his twenty-plus year career at IBM, Huang had become an expert in business transformation while helping to transform it into a global service company. The ABCDE project would help local engineers and OEM (original equipment

[24] The directive also included display technologies because the DPP government and information technology expected that displays would become as important for Taiwan as semiconductors.

[25] Lin Mou-Shiung, interviewed by Chang Kuo-Hui, November 22, 2023.

manufacturing) companies in electronics and information technology build their networks of suppliers.

Building Productive Power with China

The second pathway across the Strait—building productive power with China—gained its own burst of new energy in 2008 when the Kuomintang Party regained control of the presidency. The new KMT administration moved quickly to re-populate leadership positions with mainlanders eager to increase the density of trade and technology relations with China. They had reason to believe that the rapid expansion of shared industrial development could prove persuasive in both economic and political terms to engineers and officials on the mainland who wanted to emulate the industrial accomplishments in Taiwan. Perhaps the so-called "warming" of relations across the Strait in the near term could ground a longer-term trajectory of reconnection between Taiwan and mainland China, including possible future unification. Three thousand semiconductor engineers from Taiwan moved to China, making up 10% of all semiconductor engineers in that country (Perlez et al. 2022).

To the new government, officials who were more locally-oriented were a problem. One local engineer reported his disappointment that the incoming officials "hired directors in science and technology administration considering their color in terms of their political identities," meaning they had to be blue (KMT) rather than green (DPP). After his supervisor in the National Science and Technology Council was "ruthlessly" told to resign "even though his retirement date was only five months later," this engineer left his senior position in the Center for High-Speed Computing.[26]

In addition to encouraging the continued transfer of engineering knowledge, manufacturing plants, and personnel across the Strait, the new KMT government added a wholly new dimension: it began allowing large Taiwan companies to invest directly into China-based firms.[27] The administration did maintain the limitation on the number of Taiwan-based chip makers who could build branch foundries on the mainland. And while it began permitting the manufacturers of computer displays to build factories on the mainland, it did require them to observe the N-minus-1 policy, which was still in force.[28] But introducing the opportunity for manufacturers of semiconductors in Taiwan to become

[26] Anonymous, interviewed by Chang Kuo-Hui, 2018.

[27] This new policy began when the KMT government approved an application from TSMC to acquire ten percent of SMIC. TSMC had filed a lawsuit in the United States in 2003 alleging that SMIC had stolen its trade secrets and infringed on its patents. According to the 2009 settlement, SMIC had to pay a fine and transfer ten percent of its ownership to TSMC. The KMT government quickly approved (Clarke 2010; Reuters Staff 2009).

[28] The KMT government lifted the policy in 2011 (Ministry of Economic Affairs 2011).

co-owners of new foundries on the mainland would dramatically expand the flows of engineering knowledge and technologies from the Republic of China to the People's Republic of China. The boundary between them would most certainly become something other than a demarcation between separate countries.

Building the second pathway across the Strait also included side-trips to the United States to blur there the separation between mainlander engineers from Taiwan and engineers from China. Professional associations in the U.S. that the KMT government had supported financially began to accept membership from engineers who had traveled from China. The Chinese Institute of Engineers began recruiting members from the mainland almost immediately after the 1990s. Having long been dominated by civil engineers, the Chinese Institute directed its attention to Silicon Valley specifically to attract electronics engineers and computer scientists from China. It succeeded. By the middle of the 2000s, Chinese engineers from the mainland occupied half of the seats on its governing board.[29]

For its part, the Monte Jade Association in the United States successfully attracted a surge of interest from Chinese applicants after 2000. Since many of its members were "leaders of big companies" operating with approval from the KMT government, it sought to position itself as a kind of mediator across the Taiwan Strait by achieving parallel relations with the Chinese government. One goal was to establish one or more chapters on the mainland. A first step was to formalize a chapter in Taiwan, with a president who was a former minister of science and technology.[30] Having a chapter in Taiwan would indicate that the Monte Jade Association was not directly an agent of the Taiwan government. It furthered the point by inviting some especially well-known entrepreneurial Chinese engineers to join (Chu 2013). It also proposed to the Chinese government to establish a chapter on the mainland. Not wanting to give the impression that the Association somehow viewed China itself as part of the Chinese diaspora that it claimed to lead from the peak of Monte Jade, the Association proposed using a different Chinese name than Monte Jade Association.[31]

It is also important to note that, by this point, a certain number of Monte Jade members did include native Taiwanese who preferred to identify with multiplying ties across the Strait rather sharpening a boundary between Taiwan and China. Not surprisingly, many of them were engineer-entrepreneurs who had established working relations with factories and companies on the mainland.[32]

[29] Yang Chi-Hang Joseph (楊啟航, 1948–), Rockwell Hsu, Roger Liao and Chiu Chun-Pang, interviewed by Chang Kuo-Hui, April-June, 2018.

[30] In 2008, that minister became the first premier of the Executive Yuan.

[31] Yang Chi-Hang Joseph, interviewed by Chang Kuo-Hui, June 29, 2018. See also the Shenzhen chapter of the Monte Jade Association on the mainland (Shenzhen Mountain Jade 2023).

[32] Roger Liao, interviewed by Chang Kuo-Hui, April 3, 2018.

Centering Productive Power in China

As feared by the Lee Teng-Hui administration, expanding economic ties with companies in China could cut two ways. That country's official policy held that Taiwan remained a province of China, a position that had been affirmed in 1979 by Taiwan's loss of its seat in the United Nations. From this mainland point of view, flows of engineers and engineering knowledge from Taiwan were flows within the current boundaries of China itself. When the KMT government began allowing Taiwan's contract chip makers to invest directly in China's foundries after 2008, it was also indirectly encouraging engineers trained in Taiwan to work directly for Chinese companies. Not only were Chinese companies eager to receive them but they also had been actively recruiting engineers from Taiwan, South Korea, and Japan since the late 1990s to upgrade their own electronics and other high-tech industries as quickly as possible.

Not surprisingly, the Chinese firms preferred Taiwanese engineers on the grounds of language and culture. They did not explicitly distinguish between mainlander and native Taiwanese, however, because in their view both were provincial engineers. After the world financial crisis of 2008, municipal governments in China accelerated their efforts to attract engineers from Taiwan, using strong financial incentives. Although it is not possible to identify and trace the family histories of most of those who went, certainly some included local engineers who embraced the unification of Taiwan and China.

Chen Li-Yi (陳立宜), for example, an engineering manager at Taiwan's Chi Mei Optoelectronics Corporation, left the company in late 2009 to help Shanghai City on the mainland build a new company, Shengzhen China Star Optoelectronics. Chi Mei had long been a prominent manufacturer of liquid–crystal displays (LCDs) and was facing the prospect of having to merge with another display manufacturer to stay in business. With Chen's help, the new China Star company successfully recruited as many as 200 engineers from Taiwan. Surely, all found the financial incentives to be compelling. According to one report, an R&D engineer could receive as much as one million USD over a three-year contract, and a manager could receive one million USD per year for three years (M. Chen 2015). There is also some evidence that these engineers wanted to connect with the territories of their ancestors (M. Hsu 2000; Editorial Board 2002; Kuo and Chih 2018).

The most prominent site of active recruiting by China-based firms was through the state-owned electronics company Tsinghua Unigroup. Tsinghua Unigroup was a holding company set up by Tsinghua University on behalf of the State Council.[33] The country's thirteenth five-year plan, to cover 2016 to 2020, was going to emphasize public funding to rapidly expand the manufacture of integrated circuits. In late 2015, Inotera Memories, a Taiwan-based manufacturer of dynamic random-access memory (DRAM), approved the retirement application of its chairman Charles Kao (高啟全, 1953–). Shortly thereafter, Kao announced his move to Tsinghua Unigroup to help it manufacture memory chips.

[33] In China, universities can own companies.

Kao later asserted that his new job across the Strait complemented the old one because both companies were contributing to China. In this case, contributing to China meant recognizing a common rival in South Korea. Samsung dominated the market for DRAM at the time. "The only solution" to this dominance, Kao explained, "was to help China build its industry big enough to check and balance Samsung's power." Inotera was less able to do so within Taiwan and had in fact sought help from the U.S.-based company Micron Technology. By going to Tsinghua Unigroup from Inotera after stints at TSMC and Macronix International, Kao asserted that he was simply changing the venue for work in the service of China. His mission was to "unite Taiwan and China to defeat our most competitive rival Samsung in the global market" (Lian 2017).[34]

By 2015, it became clear that Tsinghua Unigroup sought to go beyond the manufacture of semiconductors to include their design. An initial move was to attempt to acquire MediaTek, the largest company in Taiwan dedicated to designing integrated circuits. The effort failed, however, in 2016 when the DPP regained the presidency in Taiwan and rejected the move. So Tsinghua Unigroup turned its attention to recruiting individual engineers. One of its most prominent recruits was MediaTek's Yuan Ti-Wen (袁帝文) (Cheng 2015).

The recruiting of Taiwanese engineers became active headhunting. As the owner of one design company put it, if you decline Chinese investment in your firm, "they will just come to buy out your talents" (Yen 2016). Routine recruiting by Chinese companies was so aggressive that one executive at MediaTek directed his company's telephone operators at the company to not forward calls to engineers from unknown numbers. Referring to the callers, he said, "They just randomly dial extension numbers to reach any possible engineers, and we really have few things we can do to prevent our company's engineers from talking with them" (L. Chen 2015). One engineer-entrepreneur who had declined financial entanglements with governments on the mainland was asked by municipal and provincial leaders if they were pro-independence.[35]

In the 1970s and 1980s, as explained in Chap. 4, including islander engineers alongside mainlander engineers in the emerging electronics industries had helped to animate two Taiwans, one linked to the Republic of China and the other a thing unto itself. During the 1990s and early 2000s, flows of engineers in electronics and information technology across the Strait enacted both of those Taiwans. The differences between the two collections of engineers did not blur. At the same time, those flows also contributed to

[34] Tsinghua Unigroup also recruited a senior executive from Inotera's parent company, Nanya Technology. He, like Chi Mei's Chen, then went to work recruiting engineers to his new employer and its subsidiaries (Y. Lin and H. Lin 2017). Charles Kao returned to Taiwan in 2020, citing huge debts at Tsinghua Unigroup.

[35] Anonymous, interviewed by Chang Kuo-Hui, November, 2023.

materializing a possible third Taiwan—one that reasserted its status as a province of China, but this time as the People's Republic of China.[36]

References

Bishop, John A., and Jong-Rong Chiou. 2004. "Economic Transformation and Earnings Inequality in China and Taiwan." *Journal of Asian Economics* 15(3): 549–562.

Census Office of the Executive Yuan, Taiwan 行政院戶口普查處. 1992. General Report: the 1990 Census of Population and Housing Taiwan-Fukien Area, Republic of China. Taipei: Census Office of the Executive Yuan (行政院戶口普查處).

Chang, Chun-yen 張俊彥, and Po-Lung Yu 游伯龍. 2001. "The Development of Taiwan's IC Industry: An Overview." In *Made by Taiwan: Booming in the Information Technology Era*, edited by Chun-yen Chang 張俊彥 and Po-Lung Yu 游伯龍, 3–23. River Edge, NJ: World Scientific.

Chang, Shengling, Hua-Mei Chiu, and Wen-Ling Tu. 2006. "Breaking the Silicon Silence: Voicing Health and Environmental Impacts within Taiwan's Hsinchu Science Park." In *Challenging the Chip: Labor Rights and Environmental Justice in the Global Electronics Industry*, edited by Ted Smith, David A. Sonnenfeld and David N. Pellow, 170–180. Philadelphia: Temple University Press.

Chen, Bi-Chu 陳碧珠. 2008. "Zhong xin deng tai lie cai tai ji lian dian jie bei 中芯登台獵才 台積聯電戒備 (SMIC Coming to Taiwan for Headhunting: TSMC and UMC Pay Attention)." *Jing ji ri bao* 經濟日報 *(Economic Daily)*, May 10.

Chen, Liang-Long 陳良榕. 2015. "Tai wan liang zhao chan ye de po qie wei ji 台灣兩兆產業的迫切危機 (Critical Crisis for Taiwan's IC Industry)." *Tian xia za zhi* 天下雜誌 *(CommonWealth Magazine)* 565, 118–123.

Chen, Mei-Ying 陳梅英 2015. "Xiang dang nian chen li yi tou kao hua xing 想當年…陳立宜投靠華星 (Chen Li-Yi to China Star: Taiwan's Manufacturers in Trouble)." *Zi you shi bao* 自由時報 *Liberty Times*, Mar. 24.

Cheng, Tin-Fang 鄭婷方. 2015. "Du jia zhuan fang lian fa ke qian gao ceng zhan xun gao ji fu zong cai 獨家專訪聯發科前高層、展訊高級副總裁 (Interview of MediaTek's Former Vice CEO Yuan)." *Yuan jing za zhi* 遠見雜誌 *(Global Views Monthly)* June, 89–91.

Chu, Judy 朱麗芝. 2013. "Technologist Tseng Helps Startups in China, Taiwan and the U.S.". Monte Jade Science and Technology Association of Taiwan. http://www.mjtaiwan.org.tw/pages/?Ipg=1007&showPg=1296. Accessed on May 15, 2023.

Clarke, Peter. 2010. "TSMC Applies to Take 10% Stake in SMIC." *EE Times*, Mar. 29.

Democratic Progressive Party. 1999. "DPP Resolution on Taiwan's Future." New Taiwan Ilha Formosa: The Website for Taiwan's History, Present, and Future, https://www.taiwandc.org/nws-9920.htm. Accessed on Nov. 1, 2023.

Department of Education, Taiwan Province Government 臺灣省政府教育廳, ed. 1950. *Tai wan zhi jiao yu tong ji* 臺灣之教育統計 *(Education Statistics of Taiwan Province, 1950)*. Taipei: Department of Education, Taiwan Province Government.

[36]These flows of engineers across the Strait did include local engineers, although we have not been able to determine their numbers with any confidence. At this writing, at least some local engineers have returned, citing such reasons as "the advantage of working in Taiwan is that you don't have to worry about officials shutting down the whole company because of one thought…" (Perlez et al. 2022).

Department of Statistics, Ministry of Economic Affairs, Taiwan 經濟部統計處. 2017. "Export Value Survey." Department of Stastistics, Ministry of Economic Affairs, http://dmz9.moea.gov.tw/GMWeb/common/CommonQuery.aspx. Accessed on July 31, 2023.

Department of Statistics, Ministry of Education, Taiwan 教育部統計處. 2004. *Da zhuan yuan xiao yi dong ming dan yi lan biao 1987–2003 大專院校異動名單一覽表 (1987–2003) (List of Changes in Colleges and Universities, 1987–2003)*. Taipei: Ministry of Education.

Department of Statistics, Ministry of Education, Taiwan 教育部統計處. 2006a. *Da zhuan yuan xiao gai kuang tong ji 大專院校概況統計 (General Statistics of Colleges and Universities)*. Taipei: Ministry of Education.

Department of Statistics, Ministry of Education, Taiwan 教育部統計處. 2006b. *Zhong hua min guo jiao yu tong ji 中華民國教育統計 (Education Statistics Republic of China)*. Taipei: Ministry of Education.

Department of Statistics, Ministry of Education, Taiwan 教育部統計處. 2022. *2021 Education Statistical Indicators*. Taipei: Ministry of Education.

DGBAS, Directorate General of Budget, Accounting and Statistics, Executive Yuan, Taiwan 行政院主計處, ed. 2017. *Statistics on the Number of Working Overseas in 2015*. Taipei: DGBAS.

Editorial Board, Global Views Monthly. 2002. "Zhang ru jing shang hai da zao ba cun jing yuan chang 張汝京上海打造8吋晶圓廠 (Richard Chang Builds 8-inch Wafer Fab in Shanghai)." *Yuan jian za zhi 遠見雜誌 (Global Views Monthly)*, June 1.

Education Yearbook Compilation Committee of the Ministry of Education 教育部教育年鑑編纂委員會. 1976. *Di si ci zhong hua min guo jiao yu nian jian 第四次中華民國教育年鑑 (The Fourth Republic of China Education Yearbook)*. Taipei: Zheng Zhong Shu Ju (正中書局).

Education Yearbook Compilation Committee of the Ministry of Education 教育部教育年鑑編纂委員會. 1985. *Di wu ci zhong hua min guo jiao yu nian jian 第五次中華民國教育年鑑 (The Fifth Republic of China Education Yearbook)*. Taipei: Zheng Zhong Shu Ju (正中書局).

Fuller, Douglas B. 2008. "The Cross-Strait Economic Relationship's Impact on Development in Taiwan and China: Adversaries and Partners." *Asian Survey* 48(2): 239–264.

Gao, Ting. 2003. "Ethnic Chinese Networks and International Investment: Evidence from Inward FDI in China." *Journal of Asian Economics* 14(4): 611–629.

Geppert, Linda. 2005. "Silicon Gold Rush: Taiwan's Chip Makers Can't Resist the Lure of China, the World's Largest Consumer of Semiconductors." *IEEE Spectrum* 42(6): 62–66.

Hsinchu Science Park Bureau, Ministry of Science and Technology, Taiwan 科技部新竹科學園區管理局. 2006. *Xin zhu ke xue yuan qu jiu shi wu nian bao 新竹科學園區九十五年年報 (Annual Report 2006 of Hsinch Science Park)*. Hsinch, Taiwan: Hsinchu Science Park Bureau.

Hsu, Ming-cheng 許鳴程. 2000. "Taiwan ban dao ti ren de da zhong guo qing jie 台灣半導體人的「大中國情結」(The Great Chinese Nationalism of Taiwanese Semiconductor Practitioners)." *Jin zhou kan 今周刊 (Business Today)*, June 1, 36.

Hsu, Rockwell 徐名璋. 2021. "NATEA 2.0 and Beyond." In *The North America Taiwanese Engineering & Science Association (NATEA) 30th Anniversary Special Issue: March 1991–2021*, edited by Frank Yen, Linda Huang, Rockwell Hsu, John P. Yu, Yao-hung Yang and Jesse Shiah, 19–32. Silicon Valley, CA: NATEA.

Karackattu, Joe Thomas. 2008. "The 2008 Referendum: A Test for Cross–Strait Economic Interdependence." *China Report* 44(1): 71–82.

Keating, Jerome F. 2008. *Taiwan: The Search for Identity*. Taipei: SMC Publishing Inc.

Kuo, Cheng-han 郭誠涵, and Mark Chih. 2018. "Dang da guo si wei ling jia dao guo xing ge 當「大國思維」凌駕「島國性格」(When "Big Country Thinking" Overrides "Island Country Character")." *Huan ri xian 換日線 (Crossing),* https://crossing.cw.com.tw/article/9788. Accessed on Nov. 6, 2023.

Lüthje, Boy. 2006. "The Changing Map of Global Electronics: Networks of Mass Production in the New economy." In *Challenging the Chip: Labor Rights and Environmental Justice in the Global Electronics Industry*, edited by Ted Smith, David A. Sonnenfeld and David N. Pellow, 17–30. Philadelphia: Temple University Press.

Lee, Teng-hui 李登輝. 1996. "Zong tong can jia di san jie quan guo jing ying zhe da hui 總統參加第三屆全國經營者大會 (President's Speech at the Third National Business Operators Conference)." Taipei: President Office, Taiwan.

Lee, Zong-rong 李宗榮. 2018. "Jia zu zi ben zhu yi de xing qi yu gong gu 家族資本主義的興起與鞏固 (The Rise and Consolidation of Family Capitalism in Taiwan)." In *Wei jing de qi ji : zhuan xing zhong de Taiwan jing ji yu she hui* 未竟的奇蹟: 轉型中的台灣經濟與社會 *(Unfinished Miracle: Taiwan's Economy and Society in Transition)*, edited by Zong-Rong Lee 李宗榮 and Thung-Hong Lin 林宗弘, 313–344. Taipei: Institute of Sociology, Academia Sinica.

Legislative Yuan, 立法院. 2006. "Li fa yuan gong bao wei yuan hui ji lu 立法院公報委員會紀錄 (Legislative Yuan Meeting Minutes)." Taipei: Legislative Yuan.

Leng, Tse-Kang. 2005. "State and Business in the Era of Globalization: the Case of Cross-strait Linkages in the Computer Industry." *The China Journal* (53): 63–79.

Lian, Yu-Hui 連于慧. 2017. "Jia ru zi guang ji tuan he chang jiang cun chu xin lu li cheng: Tai wan DRAM jiao fu gao qi quan du jia gao bai 加入紫光集團和長江存儲心路歷程 台灣DRAM教父高啟全獨家告白 (Story of Joining Tsinghua Unigroup: Taiwan DRAM Godfather Kao's Confession)." *DIGITIMES*, April 10.

Lin, Da-Sen 林大森. 2006. "Ji shu xue yuan yu ke ji da xue xin sheng zhu xiu ke xi zhuan huan zhi fen xi 技術學院與科技大學新生主修科系轉換之分析 (Analysis of Freshmen Major Shift in Institutes of Technology and Universities of Science and Technology)." *Jiao yu yu she hui yan jiu* 教育與社會研究 *Formosan Education and Society* (10): 93–124.

Lin, Mou-Shiung 林茂雄. 2023. *Mo er lyu cheng: Dian jing ti shu mu bao zeng de shen ci mo li* 摩爾旅程: 電晶體數目爆增的神奇魔力 *(Moore's Journey: The Magic of Exploding the Number of Transistors)*. Hsinchu, Taiwan: iCometrue.

Lin, Yuan-Ching 林苑卿, and Hong-Da Lin 林宏達. 2017. "Nian hou bai ming gong cheng shi tiao cao zhong guo nei mu 年後百名工程師跳槽中國內幕 (Inside Story of Hundred of Engineers Moving to China after Lunar New Year)." *Cai xun* 財訊 *(Wealth Magazine)*, 525: 84–88.

Mainland Affairs Council, Exectutive Yuan, Taiwan 行政院大陸委員會. 1999. "The Trade Share Ratio of Cross Strait in Taiwan's International Trade." Taipei: Mainland Affairs Council.

Mainland Affairs Council, Exectutive Yuan, Taiwan 行政院大陸委員會. 2002. "Luo shi da lu tou zi ji ji kai fang you xiao guan li zheng ce shuo ming 落實大陸投資「積極開放、有效管理」政策說明 (Explanation on the Implementation of the "Active Openness and Effective Management" Policy for Mainland China Investment)." Taipei: Mainland Affairs Council.

Mainland Affairs Council, Exectutive Yuan, Taiwan 行政院大陸委員會. 2003. "You yuan zhang dui kai fang jing yuan chang fu da lu tou zi zhi zheng ce shuo ming 游院長對開放晶圓廠赴大陸投資之政策說明 (An Explanation of Premier You's Policy on Permitting the Private Sector to Build Wafer Fabs in Mainland China in 2002)." Taipei: Mainland Affairs Council.

Miller, Chris. 2022. *Chip War: The Fight for the World's Most Critical Technology*. New York: Simon and Schuster.

Ministry of Economic Affairs, Taiwan 經濟部. 2002a. "Directions of Critical Technology Review and Monitoring Regarding Taiwan's Investment of Wafer Foundry, IC Design, IC Packaging and Testing and TFT-LCD in Mainaland China." Taipei: Ministry of Economic Affairs.

Ministry of Economic Affairs, Taiwan 經濟部. 2002b. "MOEA Special Report to Legislative Yuan: ABCDE Project." Taipei: Ministry of Economic Affairs.

Ministry of Economic Affairs, Taiwan 經濟部. 2011. "Relevant Revised Regulations for Investing in Building TFT-LCD Factories in Mainland China." Taipei: Ministry of Economic Affairs.

NATEA. 2023. "2023 U.S. Taiwan High-Tech Forum." https://www.uthf.net/. Accessed on Nov. 4, 2023.

National Applied Research Laboratories. 2023. "NARLabs Overview." National Applied Research Laboratories, https://www.narlabs.org.tw/en/xmdoc/cont?xsmsid=0I160794819089056955. Accessed on Oct. 19, 2023.

National Development Council, Executive Yuan, Taiwan 國家發展委員會. 2002. "Two Trillion and Twin Star Development Program." Taipei: National Development Council.

Perlez, Jane, Amy Chang Chien, and John Liu. 2022. "Engineers From Taiwan Bolstered China's Chip Industry. Now They're Leaving." *International New York Times,* https://www.nytimes.com/2022/11/16/business/taiwan-china-semiconductors.html. Accessed on May 27, 2024.

Rau, Dar-chin. 1998. "Transformation and Reform of Vocational Education and Training in Taiwan, Republic of China." In *Changing Vocational Education and Training: An International Comparative Perspective*, edited by Ian Finlay and Stuart Young Niven, Stephanie, 70–85. New York: Routledge.

Reuters Staff. 2009. "Long-serving SMIC Chief Richard Chang Resigns." *Reuters*, Nov. 9, 2009. https://www.reuters.com/article/idUSHKG279643/. Accessed on Nov. 4, 2023.

Sam, Choon Yin. 2007. "Economic Management in Taiwan and Mainland China: Why and How They Have been Politicized." *Global Economic Review* 36(1): 69–87.

Saxenian, AnnaLee. 2006. *The New Argonauts: Regional Advantage in a Global Economy*. Cambridge, MA: Harvard University Press.

Shenzhen Mountain Jade. 2023. "Shenzhen Mountain Jade." http://www.mjshenzhen.org/. Accessed on Oct. 23, 2023.

Smart, Alan, and Jinn-Yuh Hsu. 2004. "The Chinese Diaspora, Foreign Investment and Economic Development in China." *The Review of International Affairs* 3(4): 544–566.

Smith, Ted, David A. Sonnenfeld, and David N. Pellow. 2006. "The Quest for Sustainability and Justice in a High-tech world." In *Challenging the Chip: Labor Rights and Environmental Justice in the Global Electronics Industry*, edited by Ted Smith, David A. Sonnenfeld and David N. Pellow, 1–11. Philadelphia: Temple University Press.

Sutter, Karen M. 2002. "Business Dynamism across the Taiwan Strait: The Implications for Cross-Strait Relations." *Asian Survey* 42(3): 522–540.

Tung, Chen-Yuan 童振源, Sung-Jen Fang 方頌仁, and Wen-Hsiung Chen 陳文雄. 2016. *Xi gu cheng gong jing ji xue 矽谷成功經濟學 (Silicon Valley Success Economics)*. New Taipei: DrMaster Co., Ltd. (博誌文化).

Watanabe, Teresa. 1989. "Taiwanese 'Brains' Leave U.S.: Career Opportunities Help Lure Engineers Trained and Educated in America back Home: The trend Could Spell Trouble for U.S. High-tech Industries." *Los Angeles Times*, Dec. 29, https://www.latimes.com/archives/la-xpm-1989-12-29-mn-1114-story.html. Accessed on Oct. 31, 2023.

Wong, Seanon S. 2005. "Economic Statecraft across the Strait: Business Influence in Taiwan's Mainland policy." *Asian Perspective* 29(2): 41–72.

Wu, Chin-San 吳清山, and Hui-Ming Chien 簡慧閔. 2014. "Tai wan gao deng ji zhi jiao yu gai ge fen xi: 1997–2007 臺灣高等技職教育改革分析: 1997–2007年 (Analysis of Taiwan's Higher Technical and Vocational Education Reform)." In *Gao deng jiao yu gai ge* 高等教育改革 *(Higher Education Reform)*, edited by Chin-San Wu et al., 35–53. Taipei: Higher Education (高等教育).

Wu, Miin 吳敏求, and Chien-Rong Yang 楊倩蓉. 2022. *Wu min qiu chuan: Cong ling dao zhuo yue de zhi yu mou* 吳敏求傳：從零到卓越的識與謀 *(Wu Minn's Biography: Knowledge and Strategy from Zero to Excellence)*. Taipei: Commonwealth Publishing (天下文化).

Yen, Pei-Hua 嚴珮華. 2016. "Kai fang IC she ji ye fang huo qiang shi shen me? 開放IC設計業防火牆是什麼(What's Fire Wall for Opening IC Design Industry?)" *Tian xia za zhi* 天下雜誌 *CommonWealth Magazine* 594: 100–103.

7 To the Engineers of Taiwan, Past and Present

How you have understood yourselves and your work as engineers has been linked to how you have understood your connections to the land of Taiwan. Sometimes you have been born on that land. Other times not. In both cases, the technical work that you have carried out on that land and for its people have contributed to versions of the land in relation to other lands. That is, by means of the contents, force, and direction of your engineering work, you have helped to produce images of Taiwan that fit your own identities as persons and gained acceptance among some populations of people but not always others. Those images of Taiwan and your knowledge, identities, and commitments as engineers then further evolved together over time.

Taiwan has not always been an island, and only recently has it become a thing, or entity, unto itself. That is, it became a land that was no longer identifiable simply or purely as a subordinate part of some other land. The engineering work that you and your technicians contributed collectively over the past century and a half have played key roles in making it so. The sequence of changes in the emergence and positioning of Taiwan would not have happened without you. At the same time, Taiwan at present remains a land without a settled image of itself, an identity that has become taken for granted across it as simply true. A tension between two versions remains, and both are confronted with the possibility of a third, or even others.

Since you have had the privilege of helping Taiwan become a distinct entity in the world, you also bear responsibilities to contribute to plotting future pathways and helping to realize them. To encourage you to participate in such work and do so productively, we have focused here on helping you understand how you got to where you are.

This book traces your emergence as engineers and identifies the specific pathways through which your engineering work helped to build particular versions of Taiwan. Drawing on our own identities as STS scholars, we have mapped that emergence by examining the infrastructures of education and training that produced you as engineers and the infrastructures of work within which you have worked and lived as engineers. While you were

K.-H. Chang et al., *Engineers and the Two Taiwans*, Synthesis Lectures on Global Engineering, https://doi.org/10.1007/978-3-031-59766-4_7

learning and doing engineering, those infrastructures were pointing and challenging you. They linked your technical practices to what we have called your "geographical identity." Your geographical identity is your linkage to one place in relation to another, to a "here" in relation to a "there." The infrastructures of formation and work that you encountered signaled to you and others, at least in part, who could potentially become an engineer and what could be accomplished through engineering.

The above chapters have followed the development of infrastructures of engineering formation and work by examining five episodes of transformation and, hence, emergence. The episodes are not histories in the conventional sense. Histories tend to focus on the creators or originators of the phenomena they examine with the goal of producing comprehensive accounts of those phenomena, including what they subordinate, hide, or exclude. The episodes in this book are more limited. They explore the contents of specific relationships that emerged between dominant pathways into and through engineering and their associated attachments to the land of Taiwan. Individual engineers appear in the episodes mostly as selected examples of dominant pathways (or of resistance to them) to demonstrate and illustrate the expanse and force of those pathways and the combined technical and geographical commitments they enact.

As engineers, you learned to deal in materiality. Engaging the land of Taiwan through different pathways and making contributions through your work as engineers, you have helped to make and remake that land as distinct types of material entities.

~~~

Remember that you engineers first arrived in Taiwan in the late 1880s and early 1890s. You were not born there. You were a collection of at least eight different engineers who had been trained in the United Kingdom, Germany, Denmark, and the United States. You were hired by private companies to carry out construction projects that were under contract with a provincial government of the Qing dynasty, led by Governor Liu Ming-Chuan. In the decade or so prior to your arrival, the land of Taiwan had become a strategic military outpost for the Qing court. The Japanese navy had landed in the south in 1874 and the French navy attacked the north in 1885, calling new attention to the mainland's relationship with the land of Taiwan.

You certainly recognized the mining, telegraph, shipping, railroad, and military projects as wholly within your expertise and experience as engineers. However, you probably knew little or nothing about how you were also contributing to a self-strengthening movement that had been taking place on the mainland. That movement was seeking to rebuild the empire as a singular whole with definable borders and internal integration, a kind of entity that could compete militarily with countries and empires from the West and the Japanese archipelago.

A career soldier from the mainland, Liu worked to integrate the entire island into a singular whole, an island province of an integrated China. As a geographical entity, the island had long been a remote, partitioned territory. Fully 60% of the counties in the territory were still home to indigenous peoples and 40% were occupied by descendants
~~~

of Hakka and Han emigrants from Fujian and Guangdong provinces. The Hakka and Han peoples continued to maintain primary attachments to their home territories, treating their homesteads on the island as adjuncts to the mainland.

Liu's efforts benefitted greatly from the support of gentry merchants who had become local leaders, and he was especially interested in you engineers. While leading armies on the mainland, he had become convinced of the clear superiority of foreign weapons. He understood you and your work through the lens of *gong*, artisans who worked with their hands. To him you were a special type of *gong* who warranted elevated authority akin to gentry. He characterized you, using different characters, as master artisans (*gong shi*) who held artful work and skills, as ocean men (*yang ren*), and as ocean masters (*yang shi*). He even personally accompanied one of you on the first four miles of your survey for a new railroad.

It seems likely that you found the organization of work to be frustrating. You were not permitted to direct the technical operations that you designed and were accustomed to lead as chief engineers. You could not even supervise the soldiers who provided labor. You had to accept subordination to officials who claimed to be scholars and carried strangely different knowledge that they had presumably gained in academies and certified through exams. There was rapid turnover among you.

At least one of you actively assisted Liu in building a school to train technical personnel to work on the railroad, mines, telegraph, new arsenal, and ships. All of you likely would have recognized and endorsed his plan for the Hall of Western Learning to include daily studies of calculation, surveying, and manufacturing. Perhaps you were flattered by his desire to produce people like you, or at least prepared to support you better than untrained soldiers. You may have come to understand that people on this island seemed to value most highly some kind of classical learning and were not convinced of the value of "Western learning." In any case, they did not flock to the new school. And when Liu Ming-Chuan departed the island in 1891, official interest in engineering declined, and Liu's successor as governor closed the school. You had already left.

~~~

You started coming back sometime in the first decade of the twentieth century, but you were different. This time you had undergone formal education and training on the island of Japan in schools that drew explicitly on knowledge and practices appropriated from the United Kingdom and the empire of Germany. Your formation as engineers pointed many of you in a direction that was fairly new for the region—the construction, operation, and maintenance of machines. And rather than helping a longstanding Chinese empire avoid the risk of disintegration, you were now there to help a new, highly-integrated Japanese empire expand by building a new colony.

Shortly after the Empire of Japan had come into existence as a Meiji-led entity, it had spent much of its budget on new educational infrastructures to produce engineers mainly as knowers of machines. Machine knowledge was for industry. By building machine-based industries you would be helping to strengthen the new empire and increase its status in the
~~~

eyes of Western countries. Surprisingly for outsiders, the empire had already mobilized machine knowledge sufficiently to defeat the Qing dynasty of China and the Empire of Russia.

Along with you from the Japanese mainland came the technicians who would serve as your assistants in the colony. You likely followed your colonial leaders in understanding native men and boys to be inferior intellectually. They were incapable of achieving the levels of engineering learning that you and your technicians had mastered. Complementing your work, the colonial government established common schools for native boys and some native girls to persuade them to accept their new status as subjects of the emperor and responsibility to work on behalf of the emperor. As time passed, it also developed low-level industrial schools for native boys to prepare them for work as manual laborers in sugar refineries, chemical factories, and other industries. And it developed higher-level schools for the sons of colonists, to enable them to join you as engineers and engineering technicians.

The rapid growth of industrial production that you and others produced in the colony motivated supervisors in private industry to fill in gaps around you and your technicians with native men who had gained sufficient machine knowledge to climb above manual labor. You still enforced glass ceilings for those men, which had the effect of forcing them to choose between becoming assistant technicians with little hope of promotion or, later, departing the colony for better work and pay in Manchukuo. In any case, throughout the 50-year colonial period, you took it as given that the career pathways of natives should be subordinate to those of you who were born on the inner land or had parents who came from there.

You surely noticed that some native graduates from the industrial schools escaped your direct supervision by founding small companies. You may or may not have been aware that, prior to your arrival, many of the Han and Hakka natives in the colony had substantial experience in business. And even if you were aware, you likely didn't care if some of the native company owners characterized themselves as freelancers, claiming to be professionals and talking about working for Taiwan as if it were something other than a colony serving the emperor. You were likely happy to contract with small native companies for parts and services for your industrial machines, factories, and public construction projects and you were confident that their small companies were wholly dependent on your bigger ones.

Your contributions as engineers helped the island colony to become a stand-out contributor to the empire, bustling with industrial production and distribution across and beyond the empire. The colonial government announced that it wanted the colony to become the administrative center for a southern expansion that had already begun with an invasion of the Chinese mainland. However, as war with the Republic of China continued and then expanded to challenge the United States, you found your attention turning increasingly back to your own mainland. Some of you began to leave the colony and return home for work and service. As you left, you may or may not have been aware that native men with significant industry experience were filling in behind you. Natives who had been assistant technicians were taking on the roles and responsibilities of technicians, and some who had become technicians began to serve as engineers.

Then in the years following the end of the war virtually all of you were forced to leave, except for a few teachers and technicians who were permitted to stay on for a little while.

~~~

You came back in the late 1940s, very different once again. This time you came from mainland China and carried with you the luster of science. The War of Resistance, as the governing Kuomintang Party called it, had ended and you were there to remake the island into a province of the Republic of China. You brought an understanding of engineering as a key form of knowledge-infused practice for realizing science as moral enterprise. For roughly two decades, you had been realizing science on the mainland by advancing productivity, meaning industrial productivity. You had been helping to achieve what the KMT called national reconstruction.

As you were both reconstructing something ancient and producing something new, you were surely wholly comfortable in the understanding that your knowledge warranted high status. After all, you had studied hard to earn degrees from universities, especially the Shanghai campus of Jiao Tong University. And after graduation, you had been in leadership positions in massive technical bureaucracies, serving as technical scholar-officials responsible for industrial production.

After gaining possession of the island, your national government quickly took ownership of what had been private Japanese companies and consolidated them into companies owned by a combination of national, provincial, and local governments. Those of you who chose to move with the national government to the new province were able to transfer to new positions in the provincial government, the army, and in large industrial companies, serving as scientific specialists. The large companies that were not government-owned were still government-affiliated. They enjoyed monopoly control over their industrial arenas as well as official support for planning, design and project management. As part of the one million people who arrived by 1950, you carried the high status of mainlanders even though together you comprised roughly only a tenth of the island's population. Likely many of you hoped to reside in the island province only temporarily, until the army could successfully re-take the mainland. In any case, you could comfortably continue your work as technical scholar-officials, serving as ethical knowers for national reconstruction.

Some islander men in the new province wanted to participate in reconstruction as engineers. They said they wanted to help re-build industries that had been destroyed by American bombing. Their leaders even established new organizations to demonstrate their commitment to science. You likely agreed with provincial leaders, however, who said that islanders had been enslaved by the Japanese colonialists and that the low-level education that islanders had received had ensured their subjugation. They were not qualified to realize science as moral enterprise. You judged it appropriate that some islander men who had attended technical schools and universities in Japan could serve as engineering
~~~

technicians and engineers, but only in support positions on the peripheries of industrial production. You mainlander engineers occupied the centers.

Leaders of the new province made an effort to prepare islanders to participate in national reconstruction by reconstructing educational infrastructures in mainland terms. A key step was to introduce a distinction between national and provincial institutions. The imperial university became a national university, hence primarily for mainlanders. New primary schools for islanders replaced common schools, introducing instruction in science. Low-level secondary schools became vocational schools that taught applied science, and some industrial schools were promoted to junior colleges. One industrial school was named a provincial college. Islander families were hungry for education, and enrollments increased quickly as they willingly sent their sons and, sometimes, daughters to attend.

The ranks of engineers and engineering technicians began to include islanders by means of these new infrastructures. Those of you engineers who were islanders at the time will likely remember that, despite your preparation in the junior colleges and provincial schools, you were still judged to be subordinate. You were largely unable to gain employment in the large state-owned and state-supported companies, let alone in the provincial or national governments. As a result, after graduation many of you turned your sights wholly to the private sector. Akin to merchants and would-be engineers in previous regimes, you established small companies, this time to make consumer goods, textiles, and mechanical parts for the bigger enterprises.

Your small companies contributed to dramatic growth in the production of consumer products and led to increasing demand for islander graduates from the junior colleges and provincial school trained in mechanical and chemical engineering. The KMT government seemed content to permit, even encourage, economic initiatives by small islander companies as well as the increased production of islander technicians and engineers to work in them. The government was mostly concerned about political resistance, and its own commitment to science and scientific specialists distinguished their work from partisan politics. It was clearly confident that those of you who were mainland engineers would remain in control by virtue of your higher education and leadership in the large companies that affiliated directly with the government.

As was the case when the land of Taiwan was a colony, those of you who were islanders interested in engineering would once again have to build careers by climbing from the underside. At least this time there was a significant difference. You could do so as engineers and engineering technicians rather than solely as their assistants.

At the same time, taken together you engineers now operated with two distinct geographical identities rather than just one.

~~~

A small group of you did not fit the patterns of expectations that confronted you. You were all students in electronics engineering who had stayed in Taiwan for graduate school and elected to learn together. You considered yourselves to be abnormal and, as
~~~

such, members of a unique club. You were likely aware that those of you who were mainlanders, or perhaps now second-generation mainlanders, and those of you who were islanders would likely have significantly different career opportunities. Yet there was no clarity. So much about this new electronics industry was just taking shape. At the time, you could likely not see that your small group was modeling a practice of collaboration that would spread more generally to mainlander engineers and islander engineers working in electronics and, later, information technology.

As the small group was just getting underway, many of you who were mainlanders trained in electrical engineering were initially taking the lead in building the new industry. For the most part, you had moved to the United States for graduate school and work, often taking your families with you. Few members of your families identified with the province in any way, and most of you perhaps hoped you could return to the mainland someday to help return its land to the Republic. A combination of encouragement and pressure from the U.S. government had recently helped to persuade the KMT government to actively develop market industry in the province. It had responded by inserting itself into the private sector in an unprecedented way, directly supporting the mass manufacture of semiconductors with integrated circuits.

Developing a local electronics industry in the province meant building infrastructures to transfer knowledge and practices from abroad and reproduce them locally. Many of you who had built careers in U.S. companies actively assisted. You supported the establishment, or re-establishment, of National Chiao Tung University. You founded and organized a biennial seminar series on modern engineering and emergent technologies, with electronics a key emphasis. Some of you who had moved into government positions led the founding of a research institute to support industry, including a center focused on electronics. Over time, you played key roles in founding the first science park and in organizing ten major construction projects to support emerging industries. Some of you who were younger at the outset took advantage of opportunities to go to the United States and learn how to manufacture semiconductors. And together many of you helped to launch new companies that would operate in the private sector, albeit with substantial government support.

Meanwhile, for those of you who were islanders, the government's investment in electronics brought new opportunities for you as well. Many of you had been working in the U.S., Japanese, and European factories that the KMT government had invited to the province to produce electrical devices for export. It became clear rather quickly that electronics would not be like other industries in Taiwan built around big factories. Electronics technologies were emerging as massively distributed devices with lots of openings and niches among them. The functioning of nodes depended on the filling of openings, and relationships among them were changing fast.

You and your families founded new companies to fill those openings and help to create new niches, taking advantage of government encouragement and tolerance of small commercial enterprises. The KMT government then supported those small companies indirectly by building many new junior colleges and university programs to produce more engineers and engineering technicians to staff them. It both wanted and needed you islanders to participate in the new industry. Government officials expected to focus on the centers of power and control, the foundries, and allow small companies that would not raise political concerns to distribute the production of devices. It took for granted that the long-established hierarchical separation between state-sponsored companies controlled by mainlanders and small private companies built by islanders would continue into the new industry. As such, producing more of you would affirm the existing order, not threaten it. But your small companies multiplied rapidly, with some growing to medium-sizes. By the 1990s, you and your families were operating more than 10,000 highly-specialized producers between Hsinchu and Taipei alone.

The government's plan to invest in industry thus worked but with a significant unintended consequence. The dramatic expansion of the electronics and, later, information technology industries successfully re-made Taiwan into something entirely new, a thing on the world stage. No longer a colony or solely a small province within a big country, it was also an entity unto itself.

Those of you who participated as mainlander engineers could view this development as affirming the Republic of China, giving it a distinct transnational identity as an electronics powerhouse. But for those of you who were islander engineers, something else was happening. You were participating alongside mainlander engineers for the first time, in a wholly novel industrial infrastructure that depended upon your presence. In doing so, you were contributing materially to the emergence of something that was different than another version of China led by mainlanders. You, too, were now helping to animate Taiwan as a thing in the world. It's just that your presence was making it two things.

~~~

In Taiwan, you electronics engineers could not talk freely about any separations you noticed between mainlander and islander engineers. The ability to formally distinguish between them was realized in the United States by those of you who had moved there for education and work. You were able to name and enact the separation while avoiding actions that fellow engineers would have considered crossing a line into politics, i.e., partisan conflict. You did not embrace oppositional competition or conflict. Well, at least most of you did not.

You mainlander engineers who went first to the United States established the pattern. Your pathways justified the popular call to potential engineering students to go to Taida and then to America, meaning to find a pathway to the United States. You went to the U.S. to build your careers and potentially stay there forever. The province was, after all, basically a remote rural island, a temporary refuge.

Many of you joined professional associations that identified explicitly with China. Whether you anticipated it or not, your pursuit of learning, work, and everyday life in the U.S. likely prompted you to reassess your career attachments, your life attachments, and relationships between them. Some of you had already been members of the Chinese
~~~

Institute of Engineers. Many of you joined the Chinese-American Professionals Association and the Chinese American Academic and Professional Society. The names of these organizations attached to them a KMT version of the land of Taiwan as "Free China." Those of you who rose to leadership positions may have invited islander engineers to participate, but both of you knew that they could not follow you into leadership. Mainlanders in these organizations expected that mainlanders would lead and islanders would follow.

As the years passed, an increasing number of engineers living in the United States were you islanders. In contrast with the mainlander engineers, you tended not to bring your families, likely because of longstanding attachments to the land back home. In Taiwan, you could identify yourself publicly only with the Republic of China. In the United States, however, you had much more freedom to position yourself geographically. For the most part, the key constraint was not to advocate or appear to advocate for revolution against the KMT regime, which was still enforcing martial law. Such could have put your families at risk.

Like the mainlander engineers, you also formed associations. Initially, you tended to embrace the label "Formosa." The characters for "Taiwan" were, after all, Chinese characters. "Formosa" was recognized internationally, even though through a perspective that was clearly Western. To some of you, it was more important to resist attaching the land of Taiwan to China than to resist the memory of an attachment to Europe, especially for associations founded in the United States.

A number of you islander engineers clustered at Kansas State University, where engineering faculty and students led an informal group that explicitly discussed the politics of Taiwan. The site became a training center for independence activists. Independence activism necessarily entailed accepting partisan politics, which for engineers meant crossing a line that most avoided. Some of you identified primarily with independence activism, and had actually become engineers because it was easier to get U.S. scholarships in engineering than in the humanities or social sciences. In any case, those of you who came through Kansas State tended to become what other islander engineers called "virus carriers." You spread ideas about Taiwan identity and independence when you moved elsewhere across the country.

The rapid industrial emergence of Taiwan coupled with the waning of the KMT dream of retaking the mainland eventually helped you islander engineers justify a search for alternatives to the name Formosa. Rather than "Taiwan," some of you chose to use the adjective "Taiwanese," such as when you formed the North American Taiwanese Engineers Association. Founders of that organization shared a sense that the name had to refer to more than the land itself. It had to include people. By joining an association that called itself Taiwanese, you were explicitly attaching yourself to a land and people that were something other than China rather than positioning them as a means to China's future. Your leaders invited others to join as well, provided they considered themselves to be Taiwanese.

During the presidency of Lee Teng-Hui, to call oneself Taiwanese no longer posed a direct threat to the Republic of China because he was born in Taiwan. For the first time, the label "Taiwanese" could refer both to islanders and mainlanders. You mainlander engineers became mainland Taiwanese engineers. And you islander engineers became "native Taiwanese engineers" when in the United States or "local Taiwanese engineers" when in Taiwan.

For both of you, joining any of these professional organizations meant that you were indicating to yourselves and others that your engineering work benefited something greater than yourselves, congruent with your learned commitment to material benefit for all. You could join either "C" organizations or "T" organizations and not feel like you were betraying a broader commitment to avoid the politics of conflict. It is just that these two distinct visions of serving Taiwan contrasted with one another. Fulfilling one vision through your work in fact excluded the other. To the extent you saw yourselves as serving all was both important for and illustrative of your commitments as an engineer, but the actual effects did not match those intents. You were building different Taiwans.

The majority of you electronics engineers working in the United States did not join any of these associations. Perhaps you were prioritizing career development, seeking job stability and income for yourselves and new families that were perhaps less available back home. Perhaps organizations with different attachments looked like doing politics. But note that even while you subordinated your Taiwanese identities to your professional identities, you were still producing a parallel or overlapping outcome with those islander engineers working in Taiwan.

In the new, highly-distributed industry of electronics and, later, information technology, neither islander engineers in Taiwan nor islander engineers in the United States had to advocate publicly on behalf of Taiwan or proclaim your Taiwanese identity in order to resist continued subordination to mainlander engineers and attachments to China. You simply had to go to work, develop careers that contributed to expanding the industry, and maintain family commitments and responsibilities back home as best you could. Simply by helping to make Taiwan a thing, you were elevating islander engineers alongside mainlander engineers. You were not doing partisan politics but you were remaking power relations, elevating islanders and helping to establish the material grounds for democratization in Taiwan.

The rise of Taiwan as an entity unto itself did pose a new problem for mainland Taiwanese engineers living and working in the United States. How could you give the credit for Taiwan's increasing international visibility to the KMT government's concerted focus on the island without also severing the island's attachments to the mainland? Accepting the label Taiwanese for your associations and yourselves could risk disappearing your own mainland origins.

For some mainland Taiwanese, the solution was to identify not with Taiwan as a bounded entity but with its tallest mountain, Yushan or Monte Jade. By claiming an attachment specifically to that mountain, your new Monte Jade Association could avoid prioritizing either "Chinese" or "Taiwanese" among your members. Your privileged status as mainland Taiwanese engineers could be preserved and grow in a wholly new way, by claiming leadership of the entire Chinese diaspora that was committed to industry, especially electronics. With the name Monte Jade, you could expand the KMT vision to global terrain.

~~~

The distinction between mainland Taiwanese engineers and local Taiwanese engineers began to blur in the 1990s. Both of you were participating enthusiastically in extending technologies and companies across the Strait of Taiwan to mainland China. In both large and small/medium-sized companies and with the support of the government, you added operations on the mainland to your existing operations in Taiwan.

The actual work of transporting the production of electronics and information technologies to the Chinese mainland carried different meanings and implications for those of you with contrasting geographical identities. By traveling to the mainland and working there, you were also enacting your own understanding of the land of Taiwan in relation to the land of China, and established differences between the two main versions of that relationship would not dissolve away. As you ventured across the Strait, you traveled not as a homogeneous group but along two distinct pathways. One pathway expected to maintain the center and control of electronics production in Taiwan. The other aimed to build productive power with mainland China.

Those working specifically on the manufacture of semiconductors found yourselves having to address this tension directly. With semiconductor companies in the U.S., Europe, and sometimes at home concentrating increasingly on the design of integrated circuits, some large companies in Taiwan began shifting in the other direction toward manufacturing. Since all production operations in electronics and information technology require access to semiconductors, many of you helped these companies narrow their focus to building foundries to manufacture them, often under contract. While some companies followed the first pathway of maintaining centeredness and control in Taiwan, others took the second pathway to build productive power with and in China. By choosing one, you were rejecting the other.

During this same period, those of you who were local Taiwanese engineers found your ranks increasing dramatically in size. The government was investing heavily in junior colleges, elevating many of them to university status. As access to university degrees in engineering expanded, many local students followed the upward flow, and a high proportion sought degrees in electronics engineering, information technology, and/or computer science. Pursuing careers in electronics and information technology carried great benefits. You were both contributing to the explosion of industries that had made Taiwan a distinctive entity in the world and adding yourselves to promising expansions across the Strait into China.

Furthermore, many of you locals who had been working abroad as native Taiwanese engineers found new opportunities to return home. The termination of martial law and the
~~~

government's blacklist of dissidents offered persuasive evidence that you could embrace work in and for Taiwan without fear for yourselves or your families. In the late 1990s, the government of Lee Teng-Hui emphasized the point by inviting prominent overseas engineers to return, including native Taiwanese engineers. Returning to Taiwan enabled native Taiwanese engineers to forge new connections to the land and find new congruences between your career and family identities. Also, by returning specifically to help build industries in electronics and information technology, you could confidently view yourselves as helping the land of Taiwan to elevate its standing further, potentially into a world of countries as a member.

The growing population of local Taiwanese engineers traveling along the first pathway across the Strait gained a burst of energy when the Democratic Progressive Party (DPP) gained control of the presidency in 2000. For the first time across the land of Taiwan, local engineers gained explicit opportunities to participate in formal governance of that land. By joining the government, you could help maintain in Taiwan control over productive power in electronics. At the same time, participating in the new DPP government carried the implication of positioning yourself in opposition to the KMT party. Your material commitment to helping all as an engineer clearly became a commitment to help all through a partisan lens. You had crossed a line into partisan politics, a line that you had drawn and taken for granted.

Those of you on the second pathway across the Strait gained your own burst of energy when the KMT party regained control of the presidency in 2008. Your government moved quickly to re-populate leadership positions with mainlanders in order to increase the density of trade and technology relations with China. Those of you who joined may have hoped that a rapid expansion in shared industrial development could prove persuasive to engineers and government officials on the mainland who wanted to emulate the industrial accomplishments of Taiwan. Indeed, maybe the recent warming of official relations across the Strait could ground a longer-term trajectory of reconnection between the island and the mainland, including possible future unification. Since this goal was associated only with the KMT party, you too had to accept the fact that you had crossed a line into partisan politics.

All in all, roughly 3,000 semiconductor engineers moved across the Strait on these two pathways, along with tens of thousands of others who worked across the electronics and IT industries. While those of you who went thought you were traveling on one of two pathways, your combined flow actually did more than elevate the Taiwan that had materialized in the 1970s and 1980s. Since the Republic of China lost its seat in the United Nations in 1979, the mainland government had been asserting that Taiwan remained a part of China. When the KMT government began allowing contract chip makers to invest directly in China's foundries after 2008, it indirectly encouraged those of you on the second pathway to work directly for Chinese companies. Preferring Taiwanese engineers on the grounds of language and culture, both those companies and municipal governments accelerated their efforts to attract engineers from Taiwan. Their informal recruiting became active headhunting. As many of you went, helping the mainland to upgrade its own electronics industry and other high-tech industries, you contributed to possibly materializing a third Taiwan. That Taiwan is a land that re-asserts its status as a province of China, this time as the People's Republic of China.

~~~

In 2014, a collection of engineering professors in Taiwan joined a movement to interrupt the multiplication of material connections between mainland China and Taiwan. You were raising concerns that electronics practices from China were intruding on the routine practices of telecommunications in Taiwan and that authorizing direct Chinese investment in local telecommunications services would pose significant security risks. Even those practices that expected to maintain security control in Taiwan could eventually yield that control to the mainland.

Justified by the financial crisis of 2008, the KMT government had worked with China's government to formalize a unique type of free-trade agreement (Bureau of Foreign Trade of Ministry of Economic Affairs 2015). The draft Cross-Strait Economic Cooperation Framework Agreement (ECFA) was released in 2009 with partial implementation starting in 2010, but its Service Trade Agreement was not ratified by Taiwan's Legislative Yuan. The KMT government planned to have it effective in 2014. If ratified, it would have permitted Chinese companies to offer value-added services in telecommunications within Taiwan. The potential services included Internet access, audio conferencing, mobile virtual networking and, possibly, database and data processing. The draft agreement would, in principle, have kept in Taiwan control over the basic telecommunications infrastructure, including physical networks. However, the proposal to allow Chinese companies to participate widely in relations among Taiwanese people through such services prompted an opposition movement led by students, including many of you engineers. Both engineering students and engineering professors joined what became known as the Sunflower Movement (Ho 2015).

You professors sought to influence public opinion through petitions signed by prestigious scholars and practitioners. One petition to stop the Service Trade Agreement was signed by 289 university professors in electrical, communications, and information engineering and 249 engineers and engineering managers in information and communications industries, as well as by 134 professors in other engineering disciplines. The first three signatories were professors at Harvard University, National Taiwan University, and National Tsing Hua University, respectively. The next three were the main organizers of the professors, respectively from National Taiwan University, National Chiao Tung University, and National Cheng Kung University.

The Cheng Kung professor, Li Jung-Shian (李忠憲, ?–) wrote that the petition was the "first one in the academic history of engineers and natural sciences in Taiwan to raise a national response." Those who signed, he said, had been "known as nerds," by which he
~~~

meant that they did not participate in partisan politics. But in reviewing the draft agreement, they found that it posed a huge risk to information security, which could jeopardize Taiwan's national security in the future (Huang 2014).

For you engineers in the Sunflower Movement, likely most or all local Taiwanese engineers, allowing the Service Trade Agreement to go forward would have produced an incongruence in your identities as engineers who worked for the whole of Taiwan. You would be undermining the status of Taiwan as a distinctive thing in the world rather than elevating it. Li's references to "national" security sharpened the point that this Taiwan was a thing not only separate from China but also a national entity wholly independent from it. At the same time, those engineers who supported the agreement, likely mainland Taiwanese engineers and those local Taiwanese engineers who envisioned future unification, viewed it as multiplying connections with the mainland.

The Abnormal Club was gone. That is, the collection of engineers who had started as a small group but expanded greatly across the island, working together both directly and indirectly to help produce Taiwan as a thing and lay the material groundwork for democratization, had fallen apart. The two pathways across the Strait and the threat of a third Taiwan had made the two existing Taiwans all too visible. The differences between them were irreconcilable.

Or were they? For engineers, this question potentially has many answers, and they all come through the specific technical practices and material contributions that you develop and enact going forward as you do your work. As long as you judge your technical contributions solely in technical terms, you will prevent yourselves from recognizing that that engineering work actually does advance some version of the land of Taiwan and not others, helping some people and not others. Recognizing your work as an engineer as having political or power dimensions and implications need not be limited to crossing a line into partisan politics and the active embrace of conflict. As your engineering work actively participates in the material world, so do you.

As engineers, you have committed to provide material benefits for all. Might you be able to mobilize that commitment as a strategic choice rather than as an unrealizable goal? By means of your technical knowledge, you have played key roles in producing versions of the land of Taiwan that have always been limited. Somebody and something always got left out of your version, whether you were foreign civil engineers in the Qing province, knowers of machines in the Japanese empire, bearers of science and morality from the mainland, or members of one or the other collection of electronics engineers in an emergent Taiwan. By virtue of your technical knowledge and commitment to benefit all, you have also given yourselves responsibilities to identify who and what your engineering work includes and who and what it leaves out. Perhaps you want to leave out somebody or something through your work. That will certainly happen. But by means of your base commitment to helping everyone through your work, you also compel yourself to explain and justify to yourself (and others) how and why you might make such a choice or recognize the exclusion in the outcomes of your work. You compel yourself to ask how it might be otherwise and get to work making it so.

The Abnormal Club of electronics engineers who worked together almost in spite of themselves has gone, but what new club or clubs might you and other engineers and non-engineers form that link together people and land? The land of Taiwan is more than an island, and the people on it are many more things than local Taiwanese and mainland Taiwanese. At present, building an abnormal club of engineers that both separates and connects the lands of Taiwan and China may appear to be an impossible undertaking. But it is also impossible to imagine, let alone assess, all the engineering judgments that might be involved and what their material outcomes might include and exclude. In the meantime, might dedicated work identifying and reducing or eliminating internal inequities across the land of Taiwan produce new, even surprising, attachments and abnormal clubs? Or other potential projects? Engineering, after all, has many fields.

As engineers, your focus is of course first on your technical practices and challenges. But you can also look through and beyond those practices and challenges, starting with their links to your identities as persons. Certainly, more exclusions and more dangers lie in new pathways for engineering and engineers. So do more opportunities to attach engineering work to the land and people of Taiwan.

References

Bureau of Foreign Trade of Ministry of Economic Affairs, Taiwan 經濟部國際貿易局. 2015. "Economic Cooperation Framework Agreement." Taipei: Bureau of Foreign Trade of Ministry of Economic Affairs, Taiwan, http://www.ecfa.org.tw/. Accessed Jan. 29, 2023.

Ho, Ming-sho. 2015. "Occupy Congress in Taiwan: Political Opportunity, Threat and the Sunflower Movement." *Journal of East Asian Studies* 15(1): 69–97.

Huang, Yi-Jing 黃以敬. 2014. "Xing qi zhuan fang li zhong xian jiao shou: Kai fang zhong zi tai wan jiang qi shou dian xin wang zhan 星期專訪李忠憲教授: 開放中資, 台灣將棄守電信網戰 (Interview of Prof. Li: We Will Give up Cyberwarfare If We Are Open to Chinese Investment)." *Zi you shi bao* 自由時報 *(Liberty Times)*, May 19.

Index

K.-H. Chang et al., *Engineers and the Two Taiwans,* Synthesis Lectures on Global Engineering, https://doi.org/10.1007/978-3-031-59766-4

T

U

V

W

X

Y

Z

GPSR Compliance

The European Union's (EU) General Product Safety Regulation (GPSR) is a set of rules that requires consumer products to be safe and our obligations to ensure this.

If you have any concerns about our products, you can contact us on ProductSafety@springernature.com

In case Publisher is established outside the EU, the EU authorized representative is:

Springer Nature Customer Service Center GmbH
Europaplatz 3
69115 Heidelberg, Germany

The manufacturer's authorised representative in the EU is Springer Nature Customer Service Centre GmbH, Europaplatz 3, 69115 Heidelberg, Germany. If you have any concerns regarding our products, please contact ProductSafety@springernature.com

Printed and bound by CPI Group (UK) Ltd, Croydon, CR0 4YY
30/09/2026
02225590-0020